INTELLECTUAL PROPERTY, TECHNOLOGY TRANSFER AND GENETIC RESOURCES

AN OECD SURVEY OF CURRENT PRACTICES AND POLICIES

ORGANISATION FOR ECONOMIC CO-OPERATION AND DEVELOPMENT

ORGANISATION FOR ECONOMIC CO-OPERATION AND DEVELOPMENT

Pursuant to Article 1 of the Convention signed in Paris on 14th December 1960, and which came into force on 30th September 1961, the Organisation for Economic Co-operation and Development (OECD) shall promote policies designed:

- to achieve the highest sustainable economic growth and employment and a rising standard of living in Member countries, while maintaining financial stability, and thus to contribute to the development of the world economy;
- to contribute to sound economic expansion in Member as well as non-member countries in the process of economic development; and
- to contribute to the expansion of world trade on a multilateral, non-discriminatory basis in accordance with international obligations.

The original Member countries of the OECD are Austria, Belgium, Canada, Denmark, France, Germany, Greece, Iceland, Ireland, Italy, Luxembourg, the Netherlands, Norway, Portugal, Spain, Sweden, Switzerland, Turkey, the United Kingdom and the United States. The following countries became Members subsequently through accession at the dates indicated hereafter: Japan (28th April 1964), Finland (28th January 1969), Australia (7th June 1971), New Zealand (29th May 1973), Mexico (18th May 1994), the Czech Republic (21st December 1995) and Hungary (7th May 1996). The Commission of the European Communities takes part in the work of the OECD (Article 13 of the OECD Convention).

Publié en français sous le titre :

PROPRIÉTÉ INTELLECTUELLE, TRANSFERT DE TECHNOLOGIE ET RESSOURCES GÉNÉTIQUES
UNE ÉTUDE DE L'OCDE SUR LES PRATIQUES ET POLITIQUES ACTUELLES

FOREWORD

Intellectual Property, Technology Transfer and Genetic Resources is the first major OECD contribution to biotechnology related intellectual property issues since the 1985 publication, *Biotechnology and Patent Protection: An International Review.*

The report is the result of an activity first proposed by Switzerland in 1994, stimulated by the discussions preceding and accompanying the ratification of the Convention on Biological Diversity.

The Convention on Biological Diversity represents a political commitment to universally accepted objectives of conservation. However, it also addresses several objectives raising complex and controversial questions, including some relating to intellectual property. It is hoped that this report will make a constructive contribution to the international discussions on some of these matters.

The report is based on a questionnaire approved by the OECD's Working Party on Biotechnology (WPB) on 1-2 February 1995, country replies to that questionnaire, and other publications and information in the public domain.

Nineteen Member countries delivered replies: Australia, Austria, Belgium, Canada, Czech Republic, Finland, France, Germany, Ireland, Italy, Japan, Korea (member of CSTP), Mexico, the Netherlands, Norway, Spain, Switzerland, the United Kingdom and the United States. In addition, the European Commission provided comments.

The report was prepared by two international experts, Mr. R. Stephen Crespi (London) and Professor Joseph Straus (Munich) working in collaboration. Mr. Crespi was responsible in particular for Chapters 2, 3 and 4. They were supported by Mr. Salomon Wald and other members of the OECD Secretariat.

The text was approved by the WPB on 1-2 July 1996. The Working Party asked the OECD to ensure that the widest possible attention be drawn to the derestricted text, particularly in the relevant international fora such as the World Trade Organisation (WTO), the Secretariat of the Convention on Biological Diversity and the Food and Agriculture Organisation of the United Nations (FAO).

After incorporation of amendments submitted by Member countries, the OECD Committee for Scientific and Technological Policy (CSTP) agreed on 1-2 October 1996 to recommend the derestriction of the publication on the responsibility of the Secretary-General of the OECD.

The report does not necessarily represent the views of all OECD Member countries.

TABLE OF CONTENTS

SUMMARY OF MAIN POINTS

The report, *Intellectual Property, Technology Transfer and Genetic Resources: An OECD Survey of Current Practices and Policies,* reviews current practices and policies on intellectual property, technology transfer, and access to genetic resources, in an attempt to better understand the links between these topics that have recently been highlighted by the adoption of the Convention on Biological Diversity (CBD).

The analysis is based primarily on responses to an OECD Questionnaire which complemented work done in other parts of the Organisation in relation to the Convention on Biological Diversity. Account has also been taken of much published legal and scientific literature commentary in relation to genetic resources, including issues dealt with in this Convention. Nineteen countries have replied, which indicates considerable interest and commitment, even if the subject has been difficult for many respondents and required inter-agency, as well as public-private sector co-ordination. Responses are heterogeneous as to structure and detailed content. However, a number of common conclusions have emerged which points to the possibility of a common understanding of these complex issues.

INTELLECTUAL PROPERTY

In public discussion of intellectual property, there are often misunderstandings of its nature and limitations. Intellectual property protection cannot be extended or enforced with regard to naturally occurring or socially maintained genetic resources, although patent, trade secret and other forms of protection can be enjoyed in products or information derived from such genetic resources and in genetic resources themselves where there has been the intervention of human ingenuity and the national laws allow. This distinction between the subject matter of intellectual property – which is invariably value-added subject matter in relation to subject matter existing in the public domain – is often confused with interests in physical property or cultural behaviour. Seen in this manner, intellectual property systems would not appear to have a clear role to play with regard to resolving concerns related to ownership or use of naturally occurring or socially maintained materials or information in the public domain. Only novel products and processes, including those derived from genetic resources and genetic resources themselves, where the technical intervention of humans has achieved a result which does not occur in nature, may claim some form of intellectual property protection, for a limited time.

The availability of such intellectual property protection, in both the "home" country and that to which technology is to be transferred, is seen as a fundamental prerequisite of co-operative activities that can lead to technology transfer agreements and to foreign investment in technology-importing countries. The overriding importance of strong intellectual property laws and enforcement procedures is underlined by nearly all national responses – this is one of their most emphasized areas of agreement.

GENETIC RESOURCES

The Member countries of the OECD and many other countries continue to recognise and honour the principle of unrestricted access to genetic resources (in accordance with the International Undertaking on Plant Genetic Resources for Food and Agriculture), while also accepting the need to reconcile this with the sovereign rights of States over their own resources, and the resulting authority to determine access to genetic resources, as recognised in the Convention on Biological Diversity.

Conservation is fully consistent with the scientific exploration and technological exploitation of genetic diversity for the development of new products and processes. More than this, the application of modern science and technology to genetic resources is essential to achieving the most effective conservation of biodiversity. Conservation is also fully compatible with the principles of intellectual property protection for new products and processes.

Few of the responses identify specific projects related to the discovery or development of genetic resources. It is clear, however, that private and public agencies in some OECD Member countries (and also the European Commission) are very active in their exploration, and have accumulated considerable experience, taking into account the mutual needs and expectations of all the parties involved. So far, access to genetic resources has generally been arranged case by case. No response suggested that arrangements regarding access to and use of genetic resources have proven to be difficult to conclude or have led to unsatisfactory results for any project participants.

The principle of prior informed consent, regarding the handling and use of samples of genetic resources incident to the granting of access, is now widely accepted. Consent must be arranged between the interested parties and does not necessarily require state intervention. Universities and public R&D centres are developing and using "Material Transfer Agreements" (MTAs) which are strongly favoured by public germplasm collections. So far, consent has often been given freely, although industry expects that some sharing of benefits will be required in the future. "Codes of conduct" have been developed or are being developed.

"Access legislation" would be a more formal type of legal instrument regulating the use of genetic resources. The Biodiversity Convention provides for access to genetic resources to be subject to national legislation. So far, no OECD Member country has enacted access legislation, and only one or very few developing countries. Equally, there is no European Union legislation covering access to genetic resources in the sense of the Biodiversity Convention. No initiative towards access legislation is reported in the responses. One apparent problem would be the identification of the entity having the necessary authority to grant access, or to authorise subsequent transfer of genetic materials.

TECHNOLOGY TRANSFER

The transfer of technologies which use genetic resources, in favour of the providers of such resources, is expected to proceed on an *ad hoc* basis. No particular legal model or paradigm emerges from the responses; all traditional methods will be used, including disclosure of R&D results, licensing, information exchange, training, joint ventures, support of R&D and others.

Also, no uniform idea has emerged on how to share in a fair and equitable way the benefits arising from the use of genetic resources with the indigenous populations concerned with the conservation of those resources.

In order to give genetic resource providers due recognition and reward, various legal instruments have been mentioned. One successful example is the Letter of Collection Agreement (LOC) used by the US National Cancer Institute. A LOC is a contractual agreement providing financial reward to source countries and indigenous peoples through remuneration and technology transfer. LOCs have been negotiated with 19 developing countries.

PRIVATE SECTOR

One of the main challenges to the implementation of international genetic resource conventions is that conventions are made among governments, but it is often the private sector that seeks access to genetic resources and that creates technology that can lead to commercial and other economic benefits. The measures required to induce the private sector to transfer technology will include the creation of conditions in the recipient country that encourage voluntary technology transfer from abroad.

In spite of strong interest in principle, some specific factors may inhibit the willingness of industrial or other sectors to conclude genetic resource agreements. One factor could be high transaction costs. Another is possible underestimation of the long time-lags before R&D on a genetic resource may yield economic benefits, *e.g.* in the pharmaceutical sector. One should also consider the longer-term potential of "combinatorial" and other types of synthetic chemistry that can create and screen at low costs large numbers of diverse molecular structures of potential biological activity. Inhibitory conditions of access to the great storehouse of natural or socially maintained genetic resources could further encourage the ongoing search for chemical techniques which might reduce dependence on bioprospecting. A critical factor is whether a private sector company will be able to protect new technologies that are developed incident to a genetic resource development agreement, whether that protection comes through effective patent, plant variety, copyright or trade secret laws in the host country, or through reliance on effective contractual arrangements to protect trade secret information.

OTHER ISSUES

Economic and financial incentives and disincentives to encourage development of genetic resources can play a significant role in the decisional process preceding investment and technology transfer decisions. However, this role appears to be secondary to intellectual property protection and to the scientific and commercial assessment of genetic resource development. Marketing restrictions on products resulting from genetic resources are a significant complication for industry, but do not necessarily inhibit projects. Financial inducements to embark on projects are an encouragement but cannot be the main motivation for industry.

The responses on the main future challenges facing the legal protection of biotechnology in general, show that there are many outstanding issues which need to be discussed in international fora. However, there is no "majority opinion" on the order of importance of these challenges. Comments are wide-ranging and extend to many fundamental issues: international harmonization, restrictions of patent protection, patents considered to be "unduly broad", the interface between patents and plant variety rights, the impact of various lobbies, and others; a general question is the roles of bilateral and multilateral agreements and the combination thereof.

INTRODUCTION AND DEFINITIONS

The objective of the investigation which has led to this report is defined in the OECD Questionnaire (Annex I) as follows:

"Inform and co-ordinate the views of OECD Member countries with regard to intellectual property in relation to technology transfer by conducting a survey of current practices, experiences and expectations related to technology transfer in biotechnology, which will analyse, from a policy and economic perspective, issues of technology transfer incident to access to genetic resources."

In January 1994, Switzerland proposed to include in the OECD/DSTI Programme of Work questions of intellectual property policy. Subsequently, the proposal was submitted to the first meeting of the Working Party on Biotechnology (WPB) of the Committee for Science and Technology Policy (CSTP) on 23-24 June 1994, with the comment that the activity might focus on issues associated with biological diversity.

In response to the interest expressed by other Member countries, a revised proposal was circulated in September 1994, calling for the preparation of a questionnaire in order to define the activity more clearly. On 14-15 November 1994, a meeting of intellectual property right experts from OECD Member countries was held in Bern. The experts designed a comprehensive questionnaire, with the aim of gathering data on current practices and policies related to access to genetic resources.

This draft questionnaire was revised and approved at the second meeting of the WPB (1-2 February 1995), and sent to all Member countries (15 February 1995).

Nineteen of them delivered replies: *Australia, Austria, Belgium, Canada, the Czech Republic, Finland, France, Germany, Ireland, Italy, Japan, Korea (member of CSTP), Mexico, the Netherlands, Norway, Spain, Switzerland, the United Kingdom, the United States.* In addition, the European Commission provided comments on the draft report.

In parallel to this intellectual property right activity of the CSTP's Working Party on Biotechnology, the OECD has made another, and major, effort to assist the implementation of the Biodiversity Convention. The OECD's Expert Group on Economic Aspects of Biodiversity, under the ultimate responsibility of the Environmental Policy Committee, has recently completed a project which also began in 1994. The resulting report, *Saving Biological Diversity: Incentive Measures*, published in June 1996, examines how policy can guide human action towards the conservation and sustainable use of biodiversity, with a particular focus on the use of incentive measures. The report reviews four categories of incentives: positive incentives (monetary or non-monetary inducements); disincentives (mechanisms that internalise the costs of use of or damage to biological resources); indirect incentives (trading and other institutional arrangements); and "perverse" incentives (inducing behaviour that reduces biodiversity).

The activity of the Environment Directorate addressed Article 11 of the Convention on Biological Diversity which highlights incentive measures, and asks for "economically and socially sound measures that act as incentives for the conservation and sustainable use of components of biological diversity."

There was agreement between the Environment Directorate and the Directorate for Science, Technology and Industry that the economic incentives study would leave the discussion of intellectual property to the WPB study on access to genetic resources. Thus, it has been possible to achieve complementarity inside the OECD, and to avoid overlap.

In this report, repeated use is made of a number of key terms. Without prejudice to formal, statutory definitions found in national and international laws, the following, less formal definitions are offered to assist the general reader:

Genetic resources are understood in accordance with definitions given in the Biodiversity Convention. Thus, genetic resources are genetic materials of actual or potential value, containing functional units of heredity, and of microbial, plant, animal, or other origin. The term therefore embraces genetic materials which have been discovered, and which may already have been utilised in practical applications, as well as those which are yet to be discovered.

Technology transfer includes, but is not limited to, the disclosure of results from research and development, the licensing or assignment of intellectual property rights related to such results, exchange of information, education and training, and joint ventures. This statement is not so much a definition as an indication of some of the means by which technology transfer, as such, is achieved. The end result, in meaningful terms, is the making available to a recipient of industrial and agricultural processes and products and the relevant enabling technology for practical realisation.

Intellectual property rights are rights granted by state authority for certain products of intellectual effort and ingenuity. These rights are the subject of specific laws (statutes) enacted by parliaments or other state authority and are generally consistent with the standards outlined in the TRIPs Agreement (see

Chapter 1), which became applicable to developed countries on 1 January 1996. ***Patents*** relate to inventions. ***Designs*** relate to shapes and configurations. ***Trade marks*** relate to words or symbols applied to products or services to identify source or sponsorship. ***Plant varieties protection*** provides *sui generis* exclusive rights in plant varieties based on the UPOV model (see Chapter 2). ***Copyright*** relates to literary or artistic works and also extends to engineering drawings, computer software and other areas beyond the sphere of the arts. Apart from copyright, all the other mentioned rights must be applied for to the relevant national authority according to statutory law and procedure. ***Trade secret*** protection protects confidential ("undisclosed") information and does not require registration or formalities.

THE BACKGROUND: INTELLECTUAL PROPERTY AND GENETIC RESOURCES IN THE NEW INTERNATIONAL CONTEXT

GENETIC RESOURCES AS A SPECIAL CASE

Mineral ores, crude oil and genetic resources as the natural or social treasures of our planet have many characteristics in common, but they differ in many more. They share the characteristic of an uneven geographical spread, *i.e.* they are located in only some areas of the earth, to a varying extent in some developing countries.

However, whereas the increasing consumption and the reduced availability of crude oil, for instance, has become a matter of general concern and has enabled the source countries to accumulate wealth and gain access to modern technologies, as well as spurred the search for alternative energy sources, nothing similar can be observed with respect to genetic resources. Neither does the general public seem to be sufficiently concerned about the rapid extinction of biodiversity, nor have source countries been able to directly collect returns for the use of their genetic resources. They have, however, benefited significantly from the improved food security arising from the sharing of improvements in crops associated with unrestricted access to genetic resources.

Humanity has already obtained enormous economic benefits from the open access to genetic resources in the form of foods, medicines and industrial products; but concerns are now expressed, both about the risks of extinction, and about future conditions of access.

The main reason for the seemingly discriminatory economic treatment of genetic resources as compared with the treatment of other resources is at the same time their strength and their legal weakness: plants, animals, micro-organisms and other biological material as genetic resources are renewable resources, capable of self-replication or of being reproduced in a biological system. They perpetuate themselves thanks to the information embodied in their genetic constitutions, which they pass on to their progenies.

However, whereas individual plants, animals and other living organisms traditionally constitute private goods, the genetic information responsible for their preservation does not. The capability of self-reproduction of biological material as a carrier of genetic information clearly reveals the limits of claiming property ownership; once acquired, either legally or not, it is impossible for the original owner to prove that the genetic information used was exclusively his: *e.g.* seeds recovered for use for further propagation, genes isolated for producing transgenic animals or plants, or for producing valuable proteins through cell culture, or for the synthetic production of valuable, active biochemical substances, and the like.

This trait of self-reproduction has rendered inventions based on self-replicating biological material particularly susceptible to copying and exploitation by parties other that the innovator. Inventions having these characteristics have proven difficult to commercialise. In this regard, the availability of intellectual property protection, and in particular, patent protection, help to establish conditions that make commercialisation feasible. Intellectual property rights provide some assurances that the innovator will be able to recoup the often extensive risks and costs of developing an invention based on a biological resource.

THE NEW INTERNATIONAL CONTEXT

The Convention on Biological Diversity, signed on 5 June 1992 in Rio de Janeiro, confirmed the principle of the sovereign rights of states over their natural resources, including the authority to determine access to genetic resources by national legislation [Article 15(1), for full text, see Annex II]. For the first time, Article 1 of the Biodiversity Convention, in which the Convention's objectives are set forth, has addressed three distinct issues in a single instrument; namely, "the conservation of biological diversity, the sustainable use of its components and the fair and equitable sharing of the benefits arising out of the utilisation of genetic resources, including by appropriate access to genetic resources and by appropriate transfer of relevant technologies taking into account all rights over those resources and to technologies...".

With the coming into force of this treaty, the basic principle of sovereignty over genetic resources was confirmed. Some implications of this recognition of sovereignty are addressed in general terms in Article 15 of the Convention, which provides that access to genetic resources shall be subject to prior informed consent of the party providing the resources, unless otherwise determined by the providing party, and on mutually agreed terms. These could include contracts with regard to the sharing of commercial and other benefits arising from utilisation of such resources. One benefit contemplated, as reflected in Articles 16 and 17 of the Convention, is technology transfer. This could take the form of licenses to use proprietary technology, sharing of research and development results, or technical information or training (for full text of Articles 15, 16 and 17, see Annex II).

If patented or otherwise protected technology is at stake, such access and transfer shall be provided on terms which recognise and are consistent with the adequate and effective protection of intellectual property rights. Parties are also to "take legislative, administrative and policy measures as appropriate" aimed at securing access to and transfer of technology, both in the public domain and subject to intellectual property rights. This obligation [Article 16(3)] is subject to the mutual agreement of the parties involved in the transaction. Article 16(4) addresses actions to be taken to promote participation in technology transfer activities by the private sector. And Article 16(5), which was subject to controversial discussions in the past, calls on parties to the Convention to co-operate "subject to national and international law, to ensure that" intellectual property rights are supportive and do not run counter to the objectives of the Convention. Many developed countries have emphasized their view that the various provisions in Article 16 are complementary despite the concerns that have been expressed by parties with other views.

The Agreement on Trade-related Aspects of Intellectual Property Rights (TRIPS Agreement), signed on 15 April 1994 in Marrakech, stipulates for the first time in the history of international industrial property protection the obligation of all members of the World Trade Organization (WTO) to provide patents for both product and process inventions in all fields of technology, provided that they are new, involve an inventive step and are capable of industrial application. TRIPS allows several exceptions from this basic principle in respect of time and substantive aspects, yet at the same time it clarifies that, subject to these exceptions, patents are available and patent rights may be exercised without discrimination as to the place of invention, the field of technology and whether products are imported or locally produced. It also clearly stipulates the content of the patent right, its term of at least 20 years, and the requirements which must be fulfilled in order to allow Member States to grant compulsory licences.

The Biodiversity Convention and the TRIPS Agreement are likely to influence the future exploitation of genetic resources to the benefit of the source countries as well as of co-operating and exploiting industries. The prior informed consent necessary for the access to the resources, as the corollary of the sovereign rights of the source countries over their genetic resources, can secure to those countries a fair share in the returns from their exploitation on a contractual basis. The TRIPS Agreement provides, subject to the exclusions allowed in Article 27:1 and 27:3, that patents shall be available for any inventions, whether products or processes, in all fields of technology, provided that they are new, involve an inventive step and are capable of industrial application. This provision is relevant to inventions that derive from genetic resources and may considerably improve the prospects for their commercially successful exploitation. A number of contracts, such as those between Merck (US) and INBio from Costa Rica, or between the US Massachusetts Institute of Technology (MIT) and the Centro de Biotecnologia da Amazonia (CBA) from Brazil, or contracts between a British university-based company (Biotics Limited) with many developing countries for phytochemical screening of local flora, which were concluded in the last few years, could serve as a model for promising co-operation. They provide not only for some modest down-payments and for a fair sharing of eventual gains, but also for mandatory earmarking of a part of the returns for conservation purposes.

The following chapters will review in greater detail some of the basic concepts mentioned here and evaluate the responses of countries to the OECD Questionnaire in light of the new international context just described.

2

AN OUTLINE OF THE BASIC ISSUES

In an effort to simplify complex issues, this chapter analyses the basic concepts underlying the current international discussions on intellectual property, genetic resources and technology transfer. Chapter 1 has described the two main international legal developments which are the outcome of these discussions. The first of these, dealing with the question of access to and development of genetic resources, is the Biodiversity Convention enacted in 1992. Articles 15, 16 and 17 of this Convention, which are relevant to this investigation, are reprinted in Annex II. The second development dealing with intellectual property law arises from the conclusion in December 1993 of the Uruguay Round trade negotiations, creating the World Trade Organization (WTO) and introducing the TRIPS Agreement, which sets minimum standards of intellectual property protection. The relevant articles of the TRIPS Agreement, pertinent to this and to the other chapters, are reproduced in Annex III.

A. INTELLECTUAL PROPERTY

Forms of intellectual property relevant to biotechnology

In biotechnology, generally, patents are of principal interest, especially for the pharmaceutical and agro-biotechnology industries. To be patentable, an invention (*e.g.* a new product or process) must be *new*, involve an *inventive* step, and be *useful*, *i.e.* have an industrial application or other practical use. The patent application is officially and critically examined for these requirements. For agricultural research aimed primarily at the development of new plant varieties, plant breeders' rights (plant variety rights) are also crucially important. To obtain a plant breeder's right for a new variety (one not previously commercialised), it must be *distinct* from known varieties, *uniform* and *stable* (DUS). The variety is officially tested for these requirements.

Main types of biotechnology patent (relevant to genetic resources)

Naturally occurring substances, present as components of complex mixtures, can, in principle, be patented where they are isolated from their surroundings, identified and made available for the first time, and a process developed for producing them so that they can be put to a useful purpose. This applies to inanimate substances, as well as to living materials. In appropriate circumstances, such substances are not ruled out as mere "discoveries", but are considered as inventions by the legal authorities.

Micro-organism patents are now obtainable in most industrially developed countries, following the landmark decision of the United States Supreme Court in 1980 that the living nature of micro-organisms does not preclude them from patentability.

Plant patents are also obtainable in the United States, Europe, Japan, Australia and some other countries. To avoid legal confusion, patent law in Europe excluded plant varieties from patentability, *e.g.* in the prototype provision of the European Patent Convention (EPC), Article 53*(b)*, excludes patents for "plant and animal varieties" and "essentially biological processes for the production of plants and animals".

Animal breeds produced by traditional methods have no legal system for their protection comparable to plant breeders' rights. US patents may in principle be obtained for non-naturally occurring non-human multi-cellular living organisms, including animals. The first transgenic animal patent was issued in 1988 to Harvard University with claims covering the "onco-mouse", one in which an onco-gene has been introduced to make the animal more susceptible to cancer and therefore more sensitive in testing possible carcinogens. Transgenic animal patents are also available in European countries. A European patent was granted for the onco-mouse, but is now under formal opposition by various groups concerned with animal welfare.

Patents for chemical compounds corresponding to nucleotide sequences may also be obtained in industrially developed countries. "Gene patents" require more discussion than is feasible in this brief summary. The patent authorities in countries where such patents have been challenged consider that the gene, in its natural state, is unpatentable, but that a patent can be granted when the gene is isolated and made available for a practical, industrial or other useful purpose.

Apart from the necessity to comply with basic patent law requirements of novelty, inventiveness and industrial applicability or utility, the prospect of obtaining patent protection is also dependent on the disclosure ("technical teaching") provided in the patent application. The scope of the patent, as expressed in the patent "claims" (which are verbal definitions of the invention), is largely influenced by the character and content of the disclosure. Claims are required to be supported by, and commensurate with, the disclosure, which must enable the skilled person to put the invention to use.

In the case of an active substance derived from a microbial, plant, or other biological material source, the disclosure will identify the species of organism and, if necessary, its geographical location. If the species or strain of organism is not readily available, a deposit of the necessary biological material in a recognised culture collection may be required. The application will describe in suitable detail the method of production or extraction of the active material and the degree of purification required. The product will normally be identified by means of appropriate physical, chemical, or biological characteristics. Assigning a defined chemical structure to the compound, where possible, is considered to be the optimal form of characterisation.

In the case of an invention based on gene isolation and transfer, the methodology must also be disclosed in an enabling (repeatable) manner. The techniques of recombinant DNA technology are now well-established and can usually be adequately described for the purposes of patent applications.

Plant variety protection

Although patents for certain types of plants have been available under US law since 1930, patent law in most other countries was originally considered unsuitable for protecting new plant varieties developed by traditional breeding methods. Special national laws of plant breeders' rights (also called plant variety rights) were therefore established in the 1960s in some countries, as well as the International Union for the Protection of New Varieties of Plant (UPOV). UPOV has a current membership of 31 States. The current Convention of UPOV is that of the 1978 Act. A highly significant revision of the UPOV Agreement was concluded in 1991. Although this revision has not yet entered into force, the changes introduced through the 1991 revision have already been implemented in a number of UPOV members, and it is expected that the revision will come into force in the near future. Articles 14 and 15 of UPOV 1991 are reproduced as Annex VI.

Prior to the 1991 Act, all versions of UPOV have restricted the scope of the breeder's right primarily to the commercial marketing of the reproductive material of the protected variety. Consequently, farmers legitimately sowing seed of a protected variety are legally free to save part of the seed from the first crop of plants for sowing on their own farms to produce a second and subsequent crops (the "farm saved seed"). It was recognised that the absence of reasonable limits on the right of parties to use and sell "farm saved seed" could significantly undermine the legitimate interests of the holder of a plant breeder right. One result of this recognition was that the definition of both the plant breeders' right and the acts that constitute an infringement of that right were clarified in the 1991 revision of the UPOV. The scope of the right will henceforth subject all production/reproduction of not only propagative material but in specific circumstances the harvested material or products of the harvested material to the authorisation of the breeder. However, it is an option whether Member States include the farm saved seed in their national legislation. In practice, if farm saved seed is introduced into national legislation, royalty rates on the harvested material are expected to be lower than for bought-in seed.

Under all UPOV Conventions, breeders enjoy the so-called "breeder's privilege" or "research exemption" which gives them the freedom to use protected plant varieties in their breeding programmes to generate other (derived) varieties. Previously, this freedom extended also to the commercialisation of the derived varieties without any royalty payment to the owner of the initial variety. UPOV 1991 now extends the scope of the breeder's right also to include varieties "essentially derived" from the protected variety. A variety which is predominantly derived from the protected variety, but which does not differ in performances or value, can therefore be produced, *i.e.* bred, but cannot be commercialised without authorisation from the owner of the initial protected variety. It is important to note that a plant breeders' right under the UPOV model cannot "block" in the same fashion as in patent law the right that would be granted to a derivative variety. The concept of the extension of protection provided through the references to essentially derived varieties does not alter this basic understanding but simply clarifies the boundaries of protection granted in a particular variety.

Under the 1961 and 1978 UPOV Conventions, protection of the same entity by both plant breeders' rights and patents was forbidden (prohibition of "double protection"). This prohibition against "double protection" was not included in the UPOV Convention of 1991 as a result of the recognition that many advances in plant biotechnology and plant breeding techniques have occurred since the original UPOV treaty was developed. As a result, protection for plant innovation at the varietal level is possible through both patent and UPOV-style plant variety protection in several OECD Member countries.

The freedom of research

The freedom to carry out research is safeguarded under both patent law and plant variety rights. Under patent law, "experimental use" for research purposes is not considered to be an infringement of the rights of the patent owner. But what is purely experimental (rather than experimental for commercial purposes) is a matter of interpretation, mainly through case law, and can therefore vary according to national jurisprudence. The freedom to commercialise the resulting products of research depends on whether or not they infringe the patent claims, or are "essentially derived" or "dependent" varieties under plant variety rights. The strengthened UPOV protection will therefore go part of the way towards the strong protection given by patents. Neither system is a threat to the free use of existing or new germplasm.

The function of intellectual property

"Industrial property" systems (*e.g.* patents, trade secrets, trade marks, design protection, etc.) have been developed by states as a means for recognising and promoting innovation. Patents, for example, protect the innovator for a limited period against use of the protected subject matter (*i.e.* the patented invention) by third parties without his consent. Patent systems promote innovation by encouraging the early and effective public disclosure of inventions; patent systems universally involve publication of a full description of the invention upon grant or, in many systems, at 18 months after patent protection is originally sought. It must be clearly understood that a patent, for example, cannot hamper the free use of whatever is already in the public domain, but can only control the use by others of the inventor's novel addition to the previously existing technology. This principle of providing a temporary period of legal protection encourages the climate for innovation, to the ultimate benefit of the public as a whole. This period of protection is not yet uniform in all countries, but for patents the period is most commonly set at 20 years from the patent application date, subject to the payment of official renewal fees which rise annually. Patents also encourage investment in research and development and in the production and marketing of new products and processes.

Statutory intellectual property rights provide a basic framework for voluntary technology transfer through intellectual property right licensing, supplemented and reinforced by provisions based on the supply of know-how and other factors which may be less easy to define. Patent law demands clear definition of the protected technology and thereby establishes the scope of the rights of the innovator, identifies what is transferred to a licensee, and allows for the corresponding freedoms of third parties to be assessed.

Subject to international agreements designed to improve and unify patent protection throughout the world, a country is free to develop its own policy towards legal protection systems and legal enforceability procedures. Thus, a country is free to develop and implement measures to encourage technological innovation, technology transfer and other technology-related objectives, provided these measures conform to the minimum standards of protection mandated by the TRIPS Agreement and other multilateral treaties in the field of intellectual property law. Important factors affecting national policy will be:

- the existing level of the national technology and expectations as to its future development;
- the need to encourage technology transfer from other countries; and
- the desire to induce foreign investment in the country or region; a strong patent system is more likely to attract foreign investment.

B. GENETIC RESOURCES

Access to genetic resources: The current situation

It has long been recognised that genetic materials and the valuable substances to which they give rise in living organisms can be used for the purposes of human welfare, especially through improvements in agriculture and the nutrition and health care of humans and animals. Before their full potential can be exploited to this end, genetic resources and their ultimate expression products must be identified, isolated, and developed. For this purpose, an input from science and technology is necessary. Modern chemistry and biotechnology provide wide-ranging and sophisticated means of achieving these desirable objectives, and have already led to significant improvement in both agricultural and pharmaceutical processes and products.

The voluntary international agreement dealing with access to plant genetic resources for food and agriculture, the "International Undertaking on Plant Genetic Resources for Food and Agriculture", when first agreed, was based on the principle that plant genetic resources were the heritage of mankind. Over the years, the Undertaking was clarified through a number of interpretative annexes which *inter alia* recognised that the concept of mankind's heritage, as applied in the Undertaking, is subject to the sovereignty of the States over their plant genetic resources (Resolution 3/91).

Some public policy declarations in this context may have been based on an incomplete appreciation of the nature of intellectual property and of its limitations. Any attempt to secure intellectual property on known and available genetic resources, either in the form of raw plant germplasm or as held in public collections, would be contrary to the fundamental principle in intellectual property law which states that what is already in the public domain cannot be removed from it and privatised. This principle is universally accepted, and in patent law, is adequately covered by the general requirement for absolute novelty of the claimed invention. To make the point explicit, the 1994 revision of the *Mexican* patent law has declared unpatentable "genetic material as found in nature".

In spite of these considerations, which patent specialists have often widely publicised, the erroneous belief that agricultural materials, knowledge and skills, handed down through generations of rural communities, can somehow be expropriated, and that intellectual property systems can be used to effect such misappropriations, is often still expressed in public commentary.

Arrangements for access to genetic resources (Material Transfer Agreements)

From earliest times, the genetic resources of the richly biodiverse regions of the world have been explored in the search for crops and other species of potential economic use. Scientific interest was often present, as an adjunct to commercial interest, and has continued in its own right, spurred on with increasing momentum with the development of powerful scientific techniques for manipulating genetic material.

Access to, and exchange of, genetic material in connection with scientific research has its own justification independently of direct commercial benefit. For example, cross-country collaboration between universities and other research institutes has most often been directed to the stocking of germplasm collections and a widening of the knowledge base through scientific publication, from which the collaborating scientists benefit academically. In this vein, genetic material in public gene banks has normally been freely accessible on the basis of disclosure of information as to intended use. Similarly, gene prospecting by scientists has also depended on the goodwill of local research institutes and individuals, with little emphasis placed on monetary or other return.

In recent years, as a result of renewed interest in the screening of genetic resources for potential commercial, as well as academic, purposes, the need for more formal arrangements has been considered. Agreements prepared in industrial and academic circles for the supply and exchange of biological materials in a research context have come into use in recent years. Official public collections of germplasm are considering the use of similar agreements to cater to the needs and interests of the donors of germplasm (the provider), and those individuals and organisations (the recipient) who request material from these public sources. These agreements, termed Material Transfer Agreements (MTAs), are being drafted, or are in use, in a number of variant forms by various private or public germplasm collections. An example of an MTA under consideration at present is given in Annex IV.

MTAs are viewed by some individual commentators as affording a form of intellectual property protection. However, for reasons explained above, intellectual property in the customary legal sense does not reside in the raw natural material, and may only arise when the natural material is modified in some industrially useful manner, or a component of it is isolated and applied to a utilitarian purpose. MTAs may therefore more properly be classified as contractual agreements designed to safeguard the interest of the supplier of the material in the interim, before the generation of intellectual property, from which the supplier may obtain benefit proportionate to his contribution. MTAs frequently contain provisions covering intellectual property deriving from research on the material as supplied by the official collection. Current drafts show varying attitudes to intellectual property, and a harmonized approach may take time to achieve.

The role of International Agricultural Research Centres

The Consultative Group on International Agricultural Research (CGIAR) is an informal association of 40 public and private sector donors supporting 16 international agricultural research centres. Some of these deal with genetic resources. Issues concerning terms of access and benefit sharing of the material housed in these collections are being addressed through the revision of the International Undertaking on Plant Genetic Resources for Food and Agriculture. As part of the overall process of addressing these issues, the *ex situ* collections of plant genetic resources for food and agriculture have been formally placed under the auspices of the FAO.

The International Agricultural Research Centres' gene banks are held in trust for the world community and have been available in accordance with declared CGIAR policy. Material Transfer Agreements are in use by some, but not all, International Agricultural Research Centres. Policy with regard to intellectual property rights has so far been a matter for autonomous decision by the individual centres and some variations have necessarily resulted from this. A common provision in these Agreements is that recipients will not restrict (by means of intellectual property rights) the availability of the genetic resources "in their original form" or will not "appropriate these public goods".

In the context of a voluntary International Code of Conduct for Plant Germplasm Collections and Transfer (ICCPGGT), the United Nations Food and Agriculture Organisation (FAO) and CGIAR concluded an Agreement on 26 October 1994 whereby the plant genetic resources in CGIAR gene banks will be held in trust for the international community. This Agreement [Agreement between the (name of research centre) and the Food and Agriculture Organisation of the United Nations (FAO) Placing Collections of Plant Germplasm under the Auspices of FAO] confirmed and consolidated the current CGIAR policy of "unrestricted availability of germplasm held in their (*i.e.* International Agricultural Research Centres') gene banks" and its conservation and use "in research on behalf of the international community, in particular the developing countries". Article 3(b) of the Agreement stipulates that: "The Centre shall not claim legal ownership over the designated germplasm, nor shall it seek any intellectual property rights over that germplasm or related information." It is not clear what is embraced by the term "related information". Article 10 provides that where designated germplasm or related information is transferred to any other person or institution, the Centre shall ensure that the transferee is also bound by the above conditions.

Once again, these provisions demonstrate that the erroneous idea that intellectual property rights could restrict or appropriate public domain material is clearly hard to dispel.

For their own research results, International Agricultural Research Centres' policy has tended to be in favour of not seeking intellectual property, either for income generation or to supplement operational funding. However, the Centres have to operate in a changed research and funding environment and to collaborate with organisations for which intellectual property is a necessary counterpart to their willingness to invest in development. This has long been true of industrial organisations, and academic and public sector organisations are also now taking a more positive attitude towards protecting innovations resulting from their research. The International Agricultural Research Centres may therefore wish to review their own positions in this respect.

The MTA can be adapted to facilitate equitable collaborative research with, and development on, genetic resources in ways that recognise source-country and local community rights. The parties to the agreement will be the provider of the material, the recipient, and if necessary, the national government and the consenting local community. Depending on whether the recipient is a not-for-profit or a commercial organisation, there will be variations in the typical terms of such agreements.

In the application of MTAs in the context of plant breeding and crop improvement, the strong public sector interest and involvement has influenced the climate of opinion on these matters. The use of genetic resources in the search for new pharmaceuticals and agrochemicals ("bioprospecting") is an area where public sector involvement co-exists with a major effort on the part of industry. Initiatives have been taken in recent years to access genetic resources in the more richly biodiverse countries for the express purpose of producing valuable derived products. Arrangements have been put in place, with the consent of national governments, to enable industrial or public sector scientists from developed countries to collaborate with a local facilitator organisation in the phytochemical screening of local and often exotic plants.

Bioprospecting agreements take into account the financial needs and interests of the collaborating parties and, for this reason, tend to be much more specific than the typical MTA over benefit sharing. Payment for samples collected will be made ("up-front" payments), but the major expectations of benefit will be met by royalty payments on successful commercial exploitation of derived medical and agrochemical compounds. These desirable products lie at the end of a long, expensive and uncertain road of scientific and industrial investigation. A realistic assessment of the value of these projects to the provider organisation and nation is that the benefits are likely to be long-term.

Alternatives to traditional bioprospecting

Since the early part of this century, the search for anti-microbial and other useful substances has been approached in two main ways. One approach has been to screen natural sources such as soil samples, marine waters, insects, and tropical plants for biologically active components. In parallel, many potentially useful compounds have been synthesised chemically and screened by scientists in academia and especially in the pharmaceutical and agrochemical industries. A new method of approach is now becoming possible. Better understanding of biological mechanisms and recent advances in molecular biology have led to the identification of molecular targets underlying the pathology of many diseases. Chemical "libraries" containing millions of synthetic chemical structures which can potentially interact with biological receptors can be utilised in combination with computer-aided molecular design and the methods of solid-state chemistry to prepare compounds which interact with the target systems. This new approach is described as "combinatorial chemistry". The hope that lies in this approach is evidenced by a number of recent acquisitions and mergers involving large companies and the smaller companies that have developed this expertise. The ability of synthetic chemistry to track through the large range of chemical structures of potential biological activity revealed by these methods may considerably extend the possibility of developing new products which are biologically more effective than the "lead" compounds found in natural sources. Although nature provides an enormous storehouse of biologically active substances, the potential of which it would be unwise to write off, combinatorial chemistry may overshadow the longer process of "look-and-see" bioprospecting used in the past. This new approach has its own science-driven momentum, but will be encouraged even more if traditional bioprospecting is impeded by inhibitory conditions.

The need for conservation of genetic resources

Understanding the importance of genetic resources to mankind carries with it recognition of the need for genetic resources to be conserved for future generations. It is important to bear in mind that the scientific exploration of the microbial, plant and animal kingdoms with a view to producing innovative processes and products of potential industrial and commercial value is in no way inconsistent with, or inimical to, the conservation of genetic resources. A distinction must be made here between the investigation of microbes, plants and animals for potential pharmaceutical products, and the investigation of genetic material itself for potential gene transfer in the context of crop improvement. In the pharmaceutical context, there is the possibility that the bioactive compounds discovered, or structures related to them, may be amenable to chemical synthesis so that commercial production will not always be dependent on the original biological source material. One notable instance of this is that of natural pyrethrum and the synthetic pyrethroids which have been developed in both public institutional and industrial research, and which have captured a large part of the agricultural pesticides market. Even more striking examples are those of the original penicillins and cephalosporins of fungal origin and their semi-synthetic counterparts which are pre-eminent in modern medicine.

Where synthetic chemistry cannot completely take over from nature, it will be necessary to devise special measures to conserve plant populations. Similar considerations apply to the use of animals as sources of pharmacologically active compounds, *e.g.* in reptilian and arachnid venoms.

On the other hand, the study of plants and animals for the exploitation of their genetic components is less subject to this potential problem. Although there may be exceptions, it is more usually the case that relatively small amounts of genetic material are required for the investigation so that the original source material remains unchanged and undepleted. For example, the transfer of a gene from a wild grass to confer a valuable trait on a cultivated variety of wheat has no noticeable effect on the wild grass population. Once achieved, the process of transfer never requires repeating. This operation is therefore in no way comparable to the wholesale removal of the mineral wealth of a particular country or region.

The same compatibility with conservation must also be acknowledged for the intellectual property protection of such innovations. In view of widespread misunderstandings, it is necessary to stress again the fact that intellectual property extends only to the new inventions created from the gene that has been transferred from a sample of the natural material. There is, of course, the possibility that the new genetic combinations may, because of their advantages, supersede existing products. This consequence is entirely market-driven and depends as much on the purchasing choices made by the agricultural community as on the proprietor who has legal control of the improvement. In addition, the process of industrial competition will normally ensure that alternatives are available, including the older products which may continue to compete on price with the new ones. Intellectual property and competition laws contain checks and balances designed to prevent total control of what products are available on the market. It is particularly essential to dispel the common myth that intellectual property protection, standing alone, inevitably leads to monopoly power for the right-holder.

The Biodiversity Convention

The full text of Articles 15, 16 and 17 of the Convention is given in Annex II.

The Convention on Biological Diversity (CBD) sets out an internationally agreed policy framework for the conservation and sustainable utilisation of biological diversity, and for access to and the equitable sharing of the benefits arising from the utilisation of components of biological diversity.

Since its entry into force in December 1993, the CBD has covered all biological diversity, including animals, micro-organisms and plants. The CBD negotiations failed to resolve some access and benefit sharing issues in respect of plant genetic resources for food and agriculture. At the request of the governments which negotiated the Convention, outstanding matters concerning plant genetic resources for food and agriculture are being addressed within the FAO Global System for the Conservation and Use of Plant Genetic Resources for Food and Agriculture, in particular through the revision of the International Undertaking on Plant Genetic Resources for Food and Agriculture. In contrast to the CBD, the Undertaking is non-binding in international law. The issues concerning access and benefit sharing are complex. There are different policy and technological dimensions to the issues, depending on whether they involve, for example, plant production for food and agriculture or for pharmaceuticals.

The Convention recognises the sovereign rights of states over their natural resources, from which national governments have the authority to determine access to them. The Convention provides that, in return for allowing access to its genetic resources, a donor country may benefit through any of three mechanisms:

- participation in research;
- access to and transfer of derived technology;
- sharing in the results of research and proceeds of commercial exploitation.

Access and sharing are to be dealt with "on mutually agreed terms" and "subject to prior informed consent". Therefore, access to genetic resources must be preceded by negotiation as to the form in which benefit to the donor country is to be achieved. The Convention envisages formal arrangements for access based on the principle of prior informed consent at the official level.

Access to and transfer of technology among the contracting states is seen as necessary, both for the conservation of biological diversity, and for the use of genetic resources. Contracting parties (national governments) are given certain discretion to determine the "legislative, administrative or policy measures as appropriate" that can be taken to achieve this objective. Such measures must have the "aim that the private sector facilitates access to, joint development and transfer of technology ...for the benefit of both governmental institutions and the private sector of developing countries".

Technology transfer may be achieved by a variety of mechanisms. It will usually include the licensing of some form of proprietary right obtained either under an established statutory form of intellectual property or deriving from the possession of secret know-how and/or proprietary biological material. The Convention recognises that the technology to be transferred may be the subject of patents and other intellectual property rights. In fact, strong intellectual property systems can more effectively serve the goal of promoting private sector efforts to provide access to and transfer of technology.

COUNTRY RESPONSES AND COMMENTS ON RESPONSES

THE OECD QUESTIONNAIRE

The OECD Questionnaire, the origin of which is explained in the Introduction to this report, is set out in full in Annex I. The questionnaire was composed of three parts. Part 1 sought information on existing projects; Part 2 was directed towards an evaluation of government policies; and Part 3 addressed consideration for future policies.

Part 1 of the questionnaire was intended to obtain basic information as to arrangements that might be in place, or in prospect, for achieving a synthesis of the four principal themes of this study, namely: access to genetic resources, development of them, transfer of technology derived from such development, and the role of intellectual property in this regard.

The preamble to the questionnaire was explicit in its terms of reference to:

- "the transfer between countries of technology in the field of biotechnology related to the development of genetic resources";
- "access to genetic resources of one country by persons or organisations based in another country".

The prime focus of interest in this survey was therefore upon arrangements of trans-national character as distinct from those between parties under the same national jurisdiction. The latter types of arrangement are of interest to the extent that they reflect past experience in devising agreements which may provide models for use in trans-national situations.

The questionnaire was expressly intended for transmission to "parties having practical experience in technology transfer in biotechnology incident to access to genetic resources." Technology transfer takes many forms and is widely defined in the Background section of the questionnaire. The predominant form of technology transfer in industrially developed countries is via transfer of intellectual property rights, either of statutory form, *e.g.* patents and plant breeders' rights, or by contractual arrangements based on the supply of know-how and proprietary materials. This emphasis underpins the questionnaire as a whole.

GENERAL OBSERVATIONS ON THE RESPONSES

The following features of the responses as a whole must be noted:

1. Several of the responses indicate that, of those consulted in their own country, the low proportion replying meant that the results were not statistically significant. Consequently the overall response could not in all cases be taken as a representative national view. For example, the *French* response reported that replies were received from only 28 out of 227 organisations consulted. In other countries, however, a low response rate may also indicate that only a small proportion of the industrial and public sectors is actually engaged in the types of activities surveyed.
2. There were notable differences from country to country in the range of responding parties and in their apparent degree of commitment to the task posed by the questionnaire. From some countries, relatively few agencies responded, and the views of the private sector were far from prominently represented. Even within countries, there is a lack of homogeneity of opinion on some of the issues raised in the questionnaire, which makes a national position difficult to formulate. This is not solely due to differences between industry and the public sector.
3. As noted above, the preamble to the questionnaire had placed this question in the context of a transfer of technology between countries as distinct from technology transfer from one party to another in the same country. This was emphasized by the *French* co-ordinators of the study in a note accompanying the distribution of the questionnaire to organisations in France. In the great majority of the responses, the distinction between trans-national and other types of arrangement has not been clarified or even alluded to. This may reflect the fact that most countries have relatively little experience in framing such arrangements internationally and especially at the government level.
4. Genetic resources are of interest for agriculture and the agrochemical industry. They are also important for the health care industries as a source of medical bioactive materials which are high value products. Genetic resources, particularly microbial ones, have environmental applications important to industry and to the general public. Since the financial returns from the exploitation of genetic resources will vary with the industry, attitudes to the topics of the questionnaire among these various business interests are unlikely to be uniform. In most responses, it was difficult to discern such differences from the information provided.
5. The format of the questionnaire appeared to encourage some respondents in the mere encircling or ticking of the various options without expanding, even in the general terms requested, on the types of project or arrange-

ments made under Questions 1-4. Some responses simply tabulated figures and ratings without supplying accompanying commentary; this made it difficult to draw conclusions.

6. Some of the responses were given in general summary form rather than being sub-divided by question.

In the following sections of this report, each question is set out and a summary of the relevant responses is provided. This is followed by a commentary, in which some of the individual country responses are referred to specifically. In the light of the foregoing remarks, and particularly in view of the wide range of issues involved, it was not feasible to mention the specific standpoint of every country or contributor on every point. Where individual country responses are highlighted, it is because of the distinctive manner in which the observations were made, and does not imply that the view presented was not also shared and expressed by other countries. In addition, the commentary also includes observations which do not derive from the responses, as such, but which are based on information reaching the public domain either before, or after, the responses were received.

QUESTIONNAIRE PART 1. EXISTING PROJECTS (QUESTIONS 1-4)

Please describe the types of projects that you have sponsored, or participated in, related to the discovery or development of genetic resources:

Question 1. *Describe the participants and their roles (examples of categories of arrangements):*

a) company/company arrangements
b) university/company arrangements
c) non-profit organisation/company arrangements
d) government/company arrangements
e) government/government arrangements
f) other ______________________

The object of this question was to determine what initiatives have been taken so far to collect and develop genetic resources. The emphasis was primarily on the types of projects undertaken between countries, although it was recognised that countries with abundant genetic resources of their own would present opportunities for internal initiatives in this respect.

Responses to Question 1

Few responses identified specific types of projects. The question was clearly designed to obtain information on representative, specific projects, without requiring a comprehensive and unwieldy listing of such items. *Australian* respondents indicated in a general fashion the extensive activity undertaken to investigate Australian terrestrial and marine resources through collaboration between governmental institutions and both national and overseas organisations in both public and private sectors. These investigations were mainly aimed at the development of national agriculture, horticulture, and egg production. The *Belgian* response also showed active intervention and support for international agricultural research and problem solving. Five specific projects were identified which involved collaboration between *Belgian* universities and industry and institutions in African and Latin American countries, as well as with international bodies. These were aimed at crop protection and crop productivity improvements which would benefit agriculture in developed, as well as developing, countries. These projects were aimed at assisting farming communities in developing countries and providing low cost protein sources for the population.

The *Mexican* response identified specific research projects and crop improvement programmes on crops of national interest. These involve collaboration between national government departments, universities, private companies and national agricultural research institutions (CIMMYT, the International Agricultural Research Centre located in Mexico). Some projects are aimed at developing methods of conservation of national genetic resources through tissue culture.

In the response from *Norway*, one industrial contributor identified collaborative projects with national and foreign universities on crop protection and improvement. Similar references were present in individual contributions to the national responses from other countries (*e.g. Switzerland*), but not highlighted in the overall synthesis. The *Japanese* response referred to joint research projects (details not given) involving the government, industry, and other organisations directed to the conservation and sustainable use of genetic resources in South-East Asian countries, in partnership with governmental/semi-governmental organisations in those countries.

The *United States'* response dealt with the industrial experience separately from that of government agencies. The industrial replies were mediated though industrial associations reporting on behalf of their member companies. Several US companies have participated in projects to develop commercial products from genetic resources, but no specific details were provided. These projects were in collaboration with other companies, universities, non-profit research organisations and US and foreign government agencies (unspecified).

The US response stressed that the extensive funding by US government for research and evaluation of genetic resources is primarily science-oriented, related to human health, food and agriculture and in a non-commercial context. Four agencies were identified, two concerned with drug development and two with agriculture. For example, one agency (National Cancer Institute) has 40 years of experience in the search for anti-cancer drugs derived from plants of the African, Latin American and Asian continents, and has extended this to the search for anti-HIV compounds. Marine organisms are also extensively examined. Another agency (Fogarty International Center) had promoted International Co-operative Biodiversity Group programmes involving public and private sector participants from the United States and developing countries, including universities, non-profit research organisations, government laboratories and private companies. A key feature of these programmes is the involvement of at least one partner from the source country.

The US agricultural agencies reported on similar patterns of international collaboration. Sponsored projects include US company collaborations with developing countries for the micropropagation of banana and pineapple, and the development of pest-resistant potato, maize and cucurbits. A notable feature of the maize improvement programme is that the US company provides both the germplasm and the technology. Similar activities have been undertaken by Biotics Limited (based at Sussex University, United Kingdom) which, since 1986, and with European Commission support, has brokered the phytochemical screening of developing country flora by industrial and other specialised European research organisations. (Not mentioned in the *United Kingdom* response.)

To this particular question, the great majority of the other responses concentrated on the extent to which particular pairings of participant [*a)* to *f)*] had taken place. Almost all combinations are cited in most of the responses, the most frequently mentioned being pairings of company/company, university/company, non-profit organisation/company. Not surprisingly, the pairing that is most mentioned is that of university to company, reflecting the fact that academic biological scientists had accumulated experience and expertise in this field of research long before commercial agriculture paid attention to its potential benefits. Other possible pairings involving universities were also mentioned (by *Belgium*).

Only in a small minority of the responses (*United States*, *Australia* and *Japan*) are there indications that government-to-government arrangements exist to allow scientists to collect, characterise and store in appropriate collections (depositories) plant and animal genetic material from other countries for potential use in agriculture. Beyond stating that they exist, no description of these arrangements was given. Similar arrangements were said to exist between governments and non-governmental (international) research institutions, and also with overseas companies.

Comments

The questions in Part 1 of the Questionnaire were specific and searching, and the information on which to found responses was no doubt distributed widely throughout all countries and difficult to retrieve comprehensively. For some recipients, especially those of industry, matters of confidentiality may have been involved.

The responses to Question 1 were sufficient to show that some OECD Member countries have been considerably active in the exploration of genetic resources. Organisations in these countries are experienced in dealing with local institutions and official authorities in other countries to enable the potential of their genetic resources to be investigated. In the overall strategy of the investigations, the needs of the source country are important to the project, *e.g.* for local crop improvement, in addition to other objectives, *e.g.* for the improvement of cultivars in developed countries.

Question 2. *Describe in general terms the arrangements that you have made with regard to types of technology transfer:*

- *a) disclosure of results from research and development*
- *b) licensing or assignment of intellectual property rights related to such results*
- *c) exchange of information*
- *d) education and training*
- *e) joint ventures*
- *f) acquisition of one entity by another*
- *g) financial and other support of research activities*
- *h) other*

One purpose of identifying and developing genetic resources is to achieve technological innovation from which the provider may also benefit, *e.g.* by the transfer of derived technology. This question explored the current experiences and expectations of OECD Member countries in technology transfer through the methods by which this is commonly achieved.

Responses to Question 2

The entire context of the *United States'* response to this question was that of technology transfer between countries. The US response from the industrial sector reported that all the listed forms of technology transfer had been provided by their members but, again, no specifics were mentioned.

The *US* governmental sector emphasized the extensive technology transfer which took place incident to their participation in projects, in addition to and outside the context of intellectual property. Such would include the technical training of host country scientists, technicians and students, provision of equipment and materials, and sharing of non-proprietory information. Some agencies had experience of all the listed types of technology transfer. The Letter of Collection Agreement used by one agency (National Cancer Institute) includes provisions for technology transfer to the source country in the form of royalties and scientific exchange, but special authorisation would be necessary to license future patent rights. The same agency referred to its Co-operative Research and Development Agreements (CRADA) with host country participants, through which intellectual property rights could be transferred.

The Letter of Collection used by the National Cancer Institute is reproduced in Annex V. This is an Agreement between the NCI Division and the Source Country to investigate "the potential of natural products in drug discovery and development". The Agreement explicitly affirms the desire "to promote the conservation of biological diversity" and recognises "the need to compensate source country organisations and peoples in the event of commercialisation of a drug developed from an organism within their (*i.e.* source country) borders". The Agreement refers to "sincere efforts to transfer knowledge, expertise, and technology related to drug discovery and development to the appropriate Source Country Institution or Source Country Organization(s), subject to the provision of mutually acceptable guarantees for the protection of intellectual property associated with any patented technology."

The Letter of Collection also specifies that "Should the agent (*i.e.* from a plant collected in the Source Country) eventually be licensed to a pharmaceutical company for production and marketing ... NCI will require the successful licensee to negotiate and enter into agreements with the Source Country Government, agency, or Source Country Organization(s) as appropriate." It also addresses the concern of the Source Country Government in respect of "royalties and other forms of compensation, as appropriate". The commitment extends even to "products structurally based on the isolated natural product (*i.e.* where the natural product provides the lead for the development of invention)".

The special character of technology transfer between countries was not prominent in the responses received from the other countries. Indeed, the parties between which technology was transferred were rarely specified.

It may have been misleading to include (without explanation) disclosure of results from R&D as one of the listed categories of technology transfer since some of the contributors to the responses interpreted this as covering general publication of results, *e.g.* in the scientific literature. It is difficult to envisage any form of technology transfer between collaborating parties which does not involve disclosure of R&D results and exchange of information between the participants. This will in most instances be accompanied by the licensing or outright assignment of intellectual property arising from the projects. The responses are unanimous in citing these common types of technology transfer.

Most responses to this question were confined to giving a brief indication of the methods used or considered important in this respect. All the available methods were cited as relevant depending on the circumstances. For example, as noted in the *Japanese* response, the benefits of the project will normally be shared between the parties through disclosure of R&D results, and transfer of intellectual property in accordance with the project contract, *e.g.* by assignment of rights. Thus with projects jointly carried out by the *Japanese* government and foreign or international agencies, the resulting intellectual property rights would be transferred to one party or shared between the parties. In most cases these projects will involve exchange of information, education and training of personnel, and financial and other support of research activities. The *Italian* response (given in overall summary form) noted that different types of technology transfer tools are used according to specific targets. The *Swiss* response indicated that as far as the industrial sector is concerned, all categories of arrangements are used. For the public sector (the federal agencies dealing with technology transfer or with technical co-operation), emphasis is placed on education, training, and information exchange.

Joint ventures were less commonly cited. The *Canadian* response noted the need to avoid venture capitalists "draining small companies of their intellectual property". The *Canadian* response considered intellectual property rights licensing and assignments as essential to the orderly and effective dissemination of benefits and noted that governments, universities and not-for-profit organisations are now making increased use of these arrangements for technology transfer.

Financial and other support of research is a pre-requisite to the creation of new technology and no doubt for this reason was commonly cited. It is however not strictly speaking a method of technology transfer. Acquisitions must surely be the ultimate and extreme form of technology transfer and were rarely mentioned.

Comments

From the relatively non-specific character of the responses as a whole, one concludes that technology transfer based on genetic resources (especially between countries) is expected to proceed *ad hoc* utilising all available traditional methods and has not so far called for the devising of special models.

The responses indicate that most of the technology transfer arrangements envisaged in this question are in place in connection with actual or contemplated projects. The emphasis on particular types of transfer differs as between patents and plant variety rights and privileged access to information, depending on the industry involved.

As to benefit to source countries in terms of products, processes and enabling technology, there is little concrete information in the responses to indicate that projects are yet yielding such benefits. This underscores the conclusion that benefits are unlikely to come easily and quickly, especially from the study of medicinal plants.

Question 3. *Describe in general terms the arrangements you have made with respect to access and use of genetic resources:*

a) consent
b) compensation and other benefits
c) ownership and control of genetic material

This question assumed that access to and permission to use genetic resources would require the consent of the source, willingness to compensate the source, and some form of agreement over ownership and control of the donated material.

Responses to Question 3

Many of the answers to this question were merged with the answers to Question 4. There is little variation in the answers received. The principle of consent is universal, consent often being given freely and without expectation, as is typically the case with public gene banks. This free availability is the general rule where the material is to be used for scientific purposes. Where access is mediated through a collaborating scientist or other party in the source country the collaborator is expected to obtain consent of the relevant local authority. Universities and public research institutions are accustomed to using Material Transfer Agreements (MTAs) which are now to be found as a number of minor variants of a standard form of agreement. Public germplasm collections are developing their own versions of MTA. Industry expects that some form of agreement providing for payment will be required. As noted in the response from the *United Kingdom*, payment comes in the form of royalties on sales of products derived from the source material, the royalty level depending on whether the product is protected by intellectual property and whether the agreement is exclusive to the recipient company.

Ownership of the genetic material itself is assumed to vest in the supplier or to have been acquired by the collaborator. Individual jurisdictions of Australia have entered into research agreements which require technology transfer to other countries. Some Australian States have passed legislation providing for Crown ownership of animals and some (protected) plants, although access is generally controlled by protecting specific locations (*e.g.* national parks), and all that is found on them; protecting particular taxa (*e.g.* all indigenous mammals); and issuing licences which only allow access to specific resources, and only for a particular purpose (*e.g.* commercial or scientific purposes). The detail of the legislation varies across each jurisdiction within Australia as there is currently no nationally consistent approach to managing access to, and benefit sharing from, indigenous biological resources.

Canadian industry also expects to use written agreements defining the terms on which consent is given to use genetic resources, but material obtained from government-controlled gene banks is normally available without restriction. In the response from *France*, while consent was a common basis for access, most arrangements were based on purchase, and recompense related to use. The response from *Switzerland* indicated that, the local collaborator having once obtained local authority consent, it was normal for industry to regulate these matters by agreements which would provide appropriate methods of compensation. Several methods of compensation were enumerated, all of which could operate without dependence upon specific intellectual property considerations.

One large *Swiss* pharmaceutical and agrochemical company (not specifically mentioned in the Swiss response) has made a public statement of its policy on biodiversity prospecting. This contains commitments to preserve ecosystems and endangered species, co-operation with local researchers, provision of training and support, benefit sharing through written agreements, and the search for pragmatic solutions to open questions. A prominent *Danish* company has also declared its policy of commercialising microbial diversity, with equitable sharing of resulting benefits and scientific and commercial co-operation with source countries. In a similar policy statement, a major *British* pharmaceutical company has also stated that it will approach the acquisition of natural product source samples through Material Transfer Agreements which provide for appropriate payments to the suppliers and give them benefit for commercial exploitation. However, this would not normally involve transfer or sharing of intellectual property rights.

In addition to industrial policy statements of the kind just mentioned, scientists have also made their views known. In proposing the Manila Declaration on "The Ethical Utilization of Asian Biological Resources" (February 1992), scientists from 37 countries have declared themselves in favour of the positive aspects of the Biodiversity Convention. A "Code of Ethics for Foreign Collectors" has also been developed at the Botany 2000 Herbarium Curation Workshop held in Perth, Western Australia (April 1990).

In the *United States* response, the industrial sector confirmed that all factors have been addressed, *i.e.* consent, compensation and ownership. One government agency (Fogarty International Center) reported that all participants in projects must demonstrate informed consent in the structure of their contracts and that participants in the source country must accept a consent-to-use agreement before any use is made of collected material. As indicated above under Question 2, another agency (National Cancer Institute) included in its Letter of Collection a requirement that its licensee negotiate an agreement with the source country to provide benefits back for commercial exploitation of collected material.

Comments

In view of the connection between this and the following question, a combined comment for both Questions 3 and 4 is given under Question 4.

It should be noted that some countries have difficulties with the word "compensation", preferring the phrase "benefit sharing" as in the CBD text.

Question 4. *Describe the arrangements you have made with regard to ownership, control and protection of intellectual property rights in agreements relating to access to and use of genetic resources:*

a) patents
b) protection of undisclosed information
c) copyright
d) trade marks and service marks
e) plant variety protection
f) other

Responses to Question 4

The question was aimed at discovering what provisions would be commonly made in access agreements as regards ownership, control and protection of intellectual property. There appears to have been some misunderstanding of this question. Some countries considered it as referring to property in the resources, which led them to note that intellectual property resides not in the source material but only in what is derived therefrom, *e.g.* new plant varieties.

Some *Australian* respondents sought to distinguish between the native and the developed genetic resources, the country in which the resources were located, and the distinction between *in situ* and *ex situ* collections. An *Australian* industrial respondent emphasized that, in their experience so far, rights in the source organism remain with the collection but that rights to derived molecules are with the developer. The *Mexican* response drew the same distinction between ownership of the genetic resources and ownership of any subsequent inventions based on them.

The *Japanese* response also indicated that ownership of the genetic resource itself would not arise but that the developer of a new variety would be entitled to rights to it.

Other responses focused on ownership of the source material and indicated that the practice of retaining such ownership diverged from one institution to another.

Patents, secrecy agreements and plant variety protection are the most common types of intellectual property encountered in agreements envisaged in this question but most respondents restricted their answers to confirming this in a general way.

In the *United States* response the industry contributors confirmed that patents would be filed and the participants would pledge confidentiality in trade secrets. US government agencies require all information generated in a project to be kept secret until intellectual property has been applied for and until disclosure is mutually agreed. Reports are monitored for intellectual property right potential but oversight is applied to ensure that information flow is not impeded to the detriment of research and development. US agencies are required to seek intellectual property for inventions arising out of sponsored research but there are constraints as to ownership or sharing of legal rights. In CRADA projects (see Responses to Question 2 above) sole or joint ownership of inventions is accepted as appropriate and the US government reserves non-exclusive royalty-free licenses. The International Cooperative Biodiversity Group activities of the Fogarty International Center (see Responses to Question 1 above) are said to comply with intellectual property law and also with the Convention on Biological Diversity.

Within the US response the specific National Cancer Institute comment on this question is particularly noteworthy. This referred to the need for a "shared sense of co-operation with the source country and due consideration of the problems associated with the commercialisation of the source country's biodiversity resources". National Cancer Institute recognise that source countries and indigenous populations do not receive recognition under the strict canons of patent law, *e.g.* as to inventorship in derived intellectual property. Several alternative legal instruments are being developed to remedy this situation. The Letter of Collection (see Responses to Question 2 above) is seen as doing "what current patent law cannot".

The Letter of Collection is a contractual agreement that provides the recognition and financial reward of indigenous peoples and source countries through compensation and technology transfer, guest researcher and scientific support, access to scientific data, and benefit sharing through royalties. The response lists 19 developing countries with which Letters of Collection have been negotiated and indicates that negotiations with several more countries are underway.

Comments

The question of access to genetic resources is closely bound up with issues of ownership and control of intellectual property relating to such access. The responses to Questions 3 and 4 therefore overlap to some extent. As noted earlier, some confusion was evident as between genetic resources, as such, and the results of development of these. This apart, there was no indication that access to genetic resources was especially problematical or that it required anything other than the types of negotiation that would normally be expected in such situations, especially in a trans-national context. The Letter of Collection Agreement used by the US National Cancer Institute is clearly a valuable adjunct to intellectual property and has an important complementary legal function to perform. The subject of these questions is related also to that of Question 10 and fuller comment on it will be postponed for commentary under that Question.

QUESTIONNAIRE PART 2. EVALUATION OF GOVERNMENT POLICIES (QUESTIONS 5-10)

In this entire group of questions, recipients were asked to rate certain factors in order of importance in relation to the particular aspects of each question.

Common preamble:

Please rate the significance of each of the following factors with respect to developing agreements on research or development related to genetic resources.

(A: essential B: major importance C: minor importance D: no importance).

Questions 5, 6. *Availability of intellectual property protection (Question 5) in the country to which technology is transferred (Question 6) in your country, in the form of:*

A B C D (a) patents
A B C D (b) protection of undisclosed information
A B C D (c) copyright
A B C D (d) trade marks or service marks
A B C D (e) plant variety protection
A B C D (f) other

Question 7. *Your ability to enforce intellectual property rights effectively:*

A B C D (a) in the country to which technology is transferred
A B C D (b) in your country
A B C D (c) in other countries

Responses to Questions 5, 6 and 7

Because Questions 5, 6 and 7 were inter-related and were treated as so in most of the responses, it is convenient to take them together for purposes of analysis. The absence of legal protection for a particular product in any country which may offer a significant market opportunity leaves a loop-hole for a competitor or for local industry to exploit the product there without legal redress. Any organisation in the business of technological innovation, be it industrial or public sector, must find a way to cope with this problem.

The responses treat all three questions as related aspects of a common theme, admitting differences of emphasis rather than of substance. As noted in the *Cana-*

dian response, there is little point in obtaining intellectual property if it cannot be enforced and therefore many refrain from seeking protection in countries which lack effective enforcement procedures. In most of the responses, the availability of effective protection in the countries to which technology is transferred is rated as of the greatest importance but no-one sees this point in isolation. Protection in both home and relevant foreign country are rated as virtually of equal importance. For some, the crucial issue is the strength of their intellectual property position worldwide.

As expected from industry, patents, secrecy and plant variety rights consistently score as essential or of major importance. Strong statements in this respect came from the industrial sectoral contributions from the *United States*, *Switzerland* and many other countries. The availability of comprehensive intellectual property protection, especially patents, in both the home country and that to which technology is to be transferred, is seen as a fundamental prerequisite not only to technology transfer agreements but also to attract and maintain ancillary investment in the country receiving the technology. This is stated to be of particular importance to the smaller companies reliant on revenues from technology licensing. Plant variety protection is also seen as essential by the relevant industry. The *United States'* governmental contribution to these questions, from a non-commercial viewpoint, was consistent with that of its industrial counterpart, whilst perhaps implying that effective legal systems that govern contractual relations between parties might make up for defects in formal patent-type protection in the recipient country. The US public and private sectors both rate protection in the United States as the most essential requirement, and see strong US intellectual property protection as being instrumental in their country's success in biotechnology.

In other country responses there is some divergence between industry and the university/public sector contributors, but few now regard these matters as unimportant, which reflects the more commercially-minded approach nowadays to be found in these sources of innovation. The *Canadian* response notes that the younger research scientists are the more enthusiastic towards intellectual property. The public sector is, however, generally realistic about its difficulty in meeting the heavy cost of legal enforcement procedures without a commercial partner, and the same is true of the smaller biotechnology companies.

Comments

The dominant theme in the responses to Questions 5, 6 and 7 is the overriding importance of strong intellectual property law and enforcement procedures in the relevant countries. This requirement is stressed throughout the responses as a whole and must be assessed as a critical factor incident to any decision to embark on costly and high risk projects in this field.

Question 8. *Restrictions placed on marketing products that result from development of genetic resources:*

A B C D (a) *in the country to which technology is transferred*
A B C D (b) *in your country*
A B C D (c) *in other countries*

Responses to Question 8

All the listed factors were considered significant although the precise significance was not always explained. The question was not entirely clear to some respondents. An *Australian* respondent marked them generally as of major importance while the responses from *Austria* and the *Netherlands* considered them essential. The *Belgian* industrial sector was said to be opposed to all such restrictions in whatever country. The regulatory obstacles for the placing on the market of genetically modified organisms (GMOs) was mentioned as a particularly perceived problem. However the *Canadian* response noted that regulatory systems inevitably place restrictions on the marketing of products, the problem being to strike the right balance for all concerned. The response from *Germany* showed an uneven distribution of views notable, however, for the essential or major significance this topic has for the larger companies. The *Norwegian*, *Swiss* and *United Kingdom* responses noted the major importance of restrictions only or mainly in the home and recipient countries.

The *Japanese* response considered it essential to avoid the introduction of any unnecessary restriction on the marketing of products resulting from the use of genetic resources. The current situation may be altered depending, for example, on the outcome of the discussion on biosafety protocol now underway in connection with the Convention on Biological Diversity. The response from *Finland* also considered that no special restrictions were called for apart from those imposed on products generally. The *French* response was broadly in line with these views and noted that any such restrictions might be viewed less favourably if adopted uniformly in all countries.

The industrial sectoral contribution to the US response considered the absence of such restrictions to be essential to participation, although the potential size of the market in any country would affect their willingness to persevere in the face of restrictions. The US governmental agencies were on the whole much less influenced by these factors, although concerned on the part of their commercial partners.

Comments

Questions 8 and 9 can conveniently be taken together for commentary.

Question 9. *Financial and investment conditions, particularly:*

A B C D (a) availability of grants or subsidies for conducting research or development of genetic resources

A B C D (b) tax incentives regarding research or development of genetic resources

A B C D (c) other

Responses to Question 9

An *Australian* respondent indicated that grants and subsidies were mostly considered more important than tax incentives. However, the 150 per cent tax deductibility for industry research and development had encouraged private sector investment and the development of the national pharmaceutical industry. Funds for the large public sector breeding programmes in grains research had been essential.

These factors were rated of essential or major importance in responses from *Austria, Finland, France, Ireland, Japan, the Netherlands, Norway* and *Spain*. In the response from *Germany* these factors were especially important to small and medium-sized enterprises and public organisations. In *Belgium*, industry appreciated tax concessions while research institutions stressed the importance of grants and subsidies. One company mentioned the availability of venture capital.

The response from *Switzerland* stressed the greater significance of the general climate for private sector investment in this technology rather than fiscal intervention on the part of the state. One Swiss industrial contributor warned of unrealistic expectations of the source country, as to the level of royalties on drugs developed from genetic resources, which would militate against agreement with participants in these countries.

The *United Kingdom* response noted both the availability of grants and tax incentives as highly important. For one major company grants are an incentive if there is no obligation to share intellectual property. Tax benefits for R&D are rare in the European Union. More stress is placed in the UK response on the problem of venture capital funding and start-up cash injections.

The *United States'* governmental agencies mostly found these factors less significant to them than to commercial interests, with one exception, Fogarty International Center, which considered grants or subsidies essentially important and for which financial and investment incentives were necessary to stimulate and guide prospecting in a direction consistent with the aims of the Biodiversity Convention.

Comments

Marketing restrictions, whilst being a significant complication for the industries, will not necessarily inhibit projects for the development and exploitation of genetic resources and are not seen as more problematical in this area than for other areas of biotechnology.

Any financial inducements to embark on projects will serve as encouragement but the main motivation comes from the scientific and/or commercial assessment of their worthwhileness.

Question 10. *Conditions placed on parties prior to gaining access to genetic resources:*

A B C D (a) *requirement of "informed consent" as to possible future use of genetic resources*

A B C D (b) *requirement for disclosure of research results*

A B C D (c) *sharing, either through licenses or partial ownership interest, intellectual property rights to technology developed during the project*

A B C D (d) *other restrictions on the licensing of intellectual property rights*

A B C D (e) *profit sharing or royalty requirements*

A B C D (f) *up-front or fixed fee obligations as a condition of access to use of genetic resources*

A B C D (g) *other*

This question was the last in the series that called for ranking. The question attempted to assess the relative importance of the listed factors in the making of R&D agreements relating to genetic resources. The listed factors clearly derive from the Biodiversity Convention, which has now been ratified by most OECD Member countries. Articles 15, 16 and 17 of this Convention, which are the most relevant to this question, are given in Annex II.

The question assumed that any factor achieving a rating of, for example, "essential or major importance" would have been viewed as a positive and desirable inducement to the making of such agreements.

Responses to Question 10

Most of the factors were assessed in the higher importance categories by *Australian* respondents, presumably implying them to be a necessary feature of the relevant agreements. Once again, whether the respondents had in mind Australian or foreign genetic resources was not totally clear. For example, one Australian respondent mentioned the need for clear policies at State and Territory level regarding access to genetic resources, this presumably being a reference to the fact that responsibility for managing access to biological resources was shared by the jurisdictions making up the federation of Australia, though a majority of the responsibility falls to the States and Territories through responsibilities for land management.

In the *United States'* response, most of the US governmental contributors assessed informed consent as a necessary requirement, although one considered that it could also be a minor disincentive. The industrial contributors gauged most of the factors as very important or essential but, at the same time, as "a potential disincentive to reaching an agreement...." The industrial contributors were perhaps influenced by the element of restrictiveness implied in the various factors listed, as evidenced by the remark that "any conditions limiting the intellectual property rights of the company or other agent which assumes the financial and business risk of developing a new biomedical therapy based on genetic resources will severely curtail, if not eliminate, the prospects for investment and technology transfer to such a country".

The *Austrian, Korean* and *Spanish* responses noted most factors as very important. The *Belgian* response noted the difficulty of giving a general assessment covering a range of projects and partners of varying input and negotiating power, but mentioned most reluctance to share intellectual property rights or profits, and to a lesser degree, up-front obligations. The response from *Finland* also suggested the need for a case-by-case approach. In the *German* response, the large and smaller companies marked most factors as of high importance, as did the public-research and project-management organisations.

The *Canadian* response saw the question as implying governmental intervention in what, for the most part, are agreements between non-governmental parties, and considered that such intervention should be minimal. This response was clearly given outside the trans-national context of the Biodiversity Convention whereas the *Italian* response, while not specifically addressing this question, placed these and other issues squarely within the remit of the Convention.

The *French* response identified the need for prior informed consent and the sharing of intellectual property rights and benefits as of the greatest importance. Consent could normally be arranged between the parties and need not require state intervention. The same point was made in the *Mexican* response. The *Netherlands'* response attributed major significance to profit sharing or royalty provisions. The *Norwegian* response from the public-research institutes and ministries marked all factors as important, except up-front payments.

In the *Swiss* response, most of the listed factors are seen as highly important by the public sector but more qualified acceptance is expressed by industry. Sharing or licensing of rights is seen as of major importance by industry with one exception, which would make this dependent on the contribution made by the provider of the resources. The public sector contributors are divided on the importance of profit sharing. One industrial contributor would restrict profit sharing to the case of joint inventorship.

The *United Kingdom* response assessed most factors as essential or of major importance, but expressed a high level of concern over informed consent.

Comments

This topic has been partly addressed in connection with the Responses to Question 4 but is given more detailed treatment here.

In seeking access to the genetic resources of any country, a foreign national individual, company or institution would expect to deal with whatever authority was appropriate to confer this privilege. Where this requires recognition or remuneration of indigenous peoples for their part in preserving and enhancing the flora and fauna of their country or region, this would be broadly acceptable to the investigating entity. As indicated in some of the responses, it might be questioned whether government intervention in these matters is necessary.

However, some time has elapsed since the Responses were formulated. During this interim, the important question of access has been the subject of ongoing consideration in connection with the Biodiversity Convention. In order to give practical effect to the Sovereignty principle and to implement Article 15 of the Convention (Annex II), the possibility of access legislation arises for consideration. No OECD Member country, nor the European Union, have yet taken this step and, with one exception (The Philippines Executive Order No. 247), neither has any developing country.

In complying with the conditions of access, one apparent problem is that of identifying the official body or other entity having the necessary authority to grant access (the "Gatekeeper"). A second problem is to devise a mechanism for obtaining consent which does not involve transaction costs that would be prohibitive for the source country and inhibitory for the investigator seeking such consent. As a partial solution of this problem, it has since been suggested that patent procedure be adapted to include a requirement to demonstrate that informed consent has been obtained. According to this scheme, patent application procedure would offer a "trigger point" for checking that the necessary consent procedure has been followed. If this suggestion were considered feasible, it would presumably have to apply also to other intellectual property procedure, *e.g.* under plant variety right law.

However, the whole concept of prior informed consent is that access cannot proceed until consent has been obtained, in which case no intellectual property can have been generated. The basic problem cannot therefore be side-stepped by inserting further procedural complications into intellectual property law. There are also serious practical objections to this suggestion. First, it would be totally unprecedented in the field of intellectual property and would require substantive change in both national and international law. Secondly, the problem of identifying the gatekeeper would not be any easier for patent authorities. Thirdly, it would be seen as another point of discrimination against the biotechnology innovator and would be strongly resisted. It may also be viewed as incompatible with the anti-discriminatory provisions of TRIPS Article 27.

QUESTIONNAIRE PART 3. CONSIDERATIONS FOR FUTURE POLICIES (QUESTIONS 11 AND 12)

Question 11. *Please describe briefly measures that might be taken:*

a) to assist parties concerned with the conservation of genetic resources
b) to improve the protection of biotechnology world-wide

Question 11*a)* and *b)* address two different subjects, with Question 11*b)* being in fact more closely connected with Question 12. For this reason, the responses to Question 11*b)* and the comments thereon are dealt with under Question 12.

Responses to Question 11a)

In answer to Question 11*a)*, the *Belgian* response suggested that more public funding be provided for conservation research and support of *ex situ* collections (including microbial culture collections), whereas a key contribution to the second part would be to improve public knowledge and acceptance of intellectual property rights and biotechnology itself. *Australian* respondents considered that all parties should assist in the implementation of the Biodiversity Convention, particularly as regards *in situ* and *ex situ* measures, the ecological sustainability of genetic resource development, and more effective management and knowledge of resources through inventories and taxonomic work. The Japanese response emphasized the importance of continued support to promote R&D co-operation and various other programmes with developing countries for the conservation and sustainable use of genetic resources.

The contribution of one agency to the *United States'* response agreed that ratification of the Biodiversity Convention would be a significant contribution. Source countries should comply with the objectives of collection and use, but should be assisted in dealing with multinational corporations. Source countries should be value-adders rather than mere access-providers and should be assisted in matters of technical and business training with these ends in view. The industrial component of the US response stressed that there must be a clear and tangible incentive to conserve genetic resources. If an economic value is assigned to such resources, for example, by recognising their potential for the generation of intellectual property, there should be disincentives to their destruction by deforestation and in other ways.

The *Canadian* response called for the expansion of gene banks and the creation of international databases. The response noted that the developing countries are the major source of genetic resources and that incentives for conservation are necessary against competing domestic uses for land. The development of markets for innovation based on genetic resources, and assisted by intellectual property, would encourage conservation. The *Korean* response shared the view that developed countries should contribute assistance towards realisation of these goals.

These positive measures for encouraging conservation were noted in the response from *Spain* and also that from *France*, which proposed the creation of laboratories in developing countries which would collaborate with industry. The response from *Germany* followed similar lines and added the suggestion for grants to developing countries (a "biological diversity tax") for these purposes.

The views expressed above were echoed in the responses from other countries and supplemented by certain distinctive observations. The *Netherlands'* response added a view from industry to the effect that the problem lies mainly in the identification and selection of those resources that are of value for conservation. The response from *Switzerland* added an industrial view of the necessity to involve local indigenous communities in all such measures, to improve their education and socio-economic conditions, and reduce population pressures. One Swiss public sector agency supported the development of legislation appropriate to the needs of developing countries, in particular, the concept of "farmers' rights". However, the *Japanese* response could not support the concept of "farmers' rights" in terms of intellectual property rights. By nature, farmers' rights cannot be clearly defined, and therefore, cannot be enforced in practice. Also, the European Union and its member States have expressed the view that "farmers' rights" are a socio-economic concept which requires a precise legal and technical definition. The *Belgian* response referred to the issue of "farmers' rights" as an unusual form of "collective" right and noted the considerable practical problems involved in assessing a "fair" return and organising its appropriate distribution to indigenous communities.

One *Australian* public sector respondent remarked that policies to implement the provisions of the Biodiversity Convention should not unduly restrict access to genetic resources. For example high up-front payments for access would not assist either the utilisation or conservation of these resources. Granting exclusive bio-prospecting rights might bring short-term gains but could reduce competition and stifle innovation.

The *United Kingdom* response reported that a major company recommended the establishment of an international fund for the conservation of genetic resources and quoted a university view that biodiversity will be best promoted if it has economic value from which there is a flow-back of revenue to the resource owner. Other parts of this response are addressed mainly to the need for international consensus, at least among the most industrially developed countries, on the patentability of materials of plant or animal origin.

Question 12. *What are the most important challenges facing the legal system for protection of biotechnology:*

a) in the near term
b) in the medium term

In many responses this question was seen as closely connected with Question 11*b)*. Where appropriate therefore the relevant observations are merged in the following summary.

Responses to Question 11b) and Question 12

As in most responses, the summary of respondents from *Australia* called for an effective harmonized patent system covering both procurement and enforcement. Whilst being positive as to this, however, the response identified the need to resolve problems over the scope of biotechnology patents (the "unduly broad claims" issue). One public sector contributor saw conflict of interest problems as between large corporations and developed nations on the one hand, and the third-world reservoirs of genetic resources on the other. The summary of Australian respondents noted the lack of international consistency in the legal relationship between plant breeders' rights and patents. This is the one response in which public-sector contributors call for a legal system for the protection of animal varieties comparable with plant breeders' rights. An industrial contribution to this response called for more general recognition of the principle that intellectual property resides with those who develop inventions from genetic materials rather than those who supply the source material.

In the *United States'* response the industrial contribution summarised the problem, in both the short and long term, as one of eliminating restrictions on the availability of patent protection for biotechnology products. The public-sector contributors also called for co-ordinated systems for plant variety protection, and expressed views very similar to those of *Australian* respondents as regards patent protection. The burdensome cost of obtaining and enforcing intellectual property rights world-wide was noted, especially in the light of the many over-lapping patent claims. Adaptation to the TRIPS Agreement (see Chapter 1) was also called for.

The *Belgian* response called for improved public knowledge on intellectual property and biotechnology in all countries and the development of legal systems adapted to the needs of developing countries. Global harmonization of law was desirable, starting at the European level, extending to all OECD Member countries, the newly industrialising countries, and finally to the remaining countries. A specific separate law for the protection of biotechnological inventions was not to be recommended. Ethical questions should be directed to the exploitation rather than to the protection of inventions. The *Belgian* response also considered that the situation would be helped by the training of experts in developing countries towards a realisation of the role of intellectual property rights as aids to development. The *Canadian* response emphasized the same points but noted that a decision on whether to grant patents on higher life forms was still awaited in Canada. The *Italian* response referred to ongoing European attempts to harmonize patent law in this field.

The response from *France* identified four themes relevant to this question in the short term. One is the recurring idea of legislation specific for biotechnology, which has its supporters and its opponents. The need for world-wide extension of systems for plant variety protection is another consistent theme. Third is the fear of appropriation of genetic resources. Finally, attention is drawn to the difficulty of apportioning value to the respective contributions made toward the exploitation of genetic resources. In the longer term the abolition of the distinction between patents and plant variety rights could be contemplated. The fear that excessive financial demands of source countries may stifle research into the development of genetic resources is seen as a significant factor for the smaller biotechnology companies. The *German* response notes the major problems of harmonization and public acceptance of patenting, and the adaptation to TRIPS and the Biodiversity Convention. Reconciling the need to encourage the private sector and ensure third world access to techniques and products is seen as a major challenge.

The *Netherlands'* response also supports the development and harmonization of patent law on the basis of the traditional established criteria of novelty, inventive step and industrial applicability (utility). Unclear and unfounded restrictions on patentability should be removed, and greater clarity is necessary on patent scope and exhaustion of rights in relation to living material. Ethical considerations and the concerns of special interest groups should be dealt with outside patent legislation.

The *Norwegian* response was somewhat indefinite as to its own national position on this question but noted the standpoint of developing countries which have traditionally been against strong legal protection systems.

The *Swiss* response called for the wider acceptance of biotechnology and a more harmonized and positive acceptance of the need for equal treatment of this technology under intellectual property law. On some points the response from Switzerland showed an element of polarisation between one part of the public sector and industry. The patentability of plants is one area of such disagreement. For example, for industry, patents for plants and animals are indispensable. The industry view is that efforts to promote public understanding and acceptance of the legitimate use of these legal instruments is also necessary at the political level.

The *United Kingdom* response to this question listed a range of views from various contributors which cannot conveniently be synthesised into a single theme: the gene patenting issue; harmony on patentability and scope; and the activities of anti-biotechnology patent lobbies.

A warning note concerning intellectual property law is sounded in the response from *Korea*. Innovation in biotechnology is seen as a most important potential contribution to the improvement of human life in the next century. To achieve this, technology transfer has to spread to all countries without excessively high cost. The strengthening of intellectual property law should not erect barriers between developed and developing countries but should be more flexible and morally orientated if it is to achieve these objectives.

Comments

The wide-ranging responses to this question, although given in the specific context of the questionnaire, extended to many fundamental issues of general concern to intellectual property affecting biotechnology as a whole. These pertain to the problems of international harmonization, restrictions on patent protection for certain types of product, the paradoxical issue of unduly broad patents in some jurisdictions, and the interface between patents and plant variety rights, which is still confused in some countries. Perhaps there is also a need to counter the effects of the anti-biotechnology lobbies on public opinion and to promote greater understanding of biotechnology and the role of legal protection in bringing its benefits to society.

CONCLUSIONS

At this early stage in the international discussion of these issues, the demanding nature of this investigation helps to explain the incompleteness of the response from many OECD Member countries. It appears that more investigation would be necessary to do justice to the standpoints of all the key potential contributors to this debate, especially from the industrial sector. There may be divergence of opinion and attitude within both the public as well as the private sectors which remain to be resolved in many countries. The conclusions which can be drawn at this stage, therefore, while firm and clear, could well be modified in the light of further information.

Whether or not linkages can be made between access to genetic resources, technology transfer and intellectual property has been a topic of keen debate for some years now, but has been brought into more definite focus by the Convention on Biological Diversity and its implementation. In requesting information from OECD Member countries, the questionnaire was aimed at determining whether any models had been developed by or in Member countries which relate conditions of access to genetic resources to a commitment to return to the source country a share in the benefit gained by the recipient as a result of the development of these resources. The main type of benefit addressed in the questionnaire was that resulting from improved industrial or agricultural technology.

The responses almost invariably state that all the factors listed in the questionnaire have been addressed by governments and by public- and private-sector organisations within their countries. However, in view of the relatively short time since the Biodiversity Convention was adopted, it is not surprising that few instances, if any, have been described demonstrating concrete results from the development of genetic resources which have already enabled donor countries to be "compensated" in a formalized manner, either by monetary payments or by enrichment of their own national technology through technology transfer. The great achievements of the past through plant breeding to make improved cultivars available to developed and developing countries have not required formalized procedures. The public-sector effort in this direction has made these improvements freely available to agricultural communities (the "Green Revolution") and the commercial breeders have marketed the proprietary products they have themselves developed.

There is no indication in the responses that any OECD Member country is at present developing access legislation which could form a model for implementing the Biodiversity Convention on these issues. If developing countries decide to introduce access legislation, considerable expertise will be necessary to address constructively the interests of all parties involved, including indigenous communities. Measures which involve high transaction costs would almost certainly be counterproductive, not only to the interests of the parties but also to the conservation of genetic resources. Voluntary agreements and the establishement of codes of practice which commend themselves to prospectors seem to be the best way forward for the present. Such voluntary agreements and the establishment of codes of practices would also be not detrimental to the adequate and effective implementation of intellectual property law by the relevant adminstrative bodies.

In the area of crop improvement, there is already an abundance of germplasm for plant breeders to exploit, either in the public collections or in commercially available breeders' lines from which desirable traits can be transferred. It is only through realistic estimates of the demand for genetic resources, further characterisation of stored germplasm, and correspondingly realistic conditions of access, that the objectives of the Convention will be achieved.

Most respondents found Part 1 of the Questionnaire the most searching and the most difficult to handle in a general way. No paradigm appears yet to exist for the coupling together in a legal framework of access to genetic resources and technology transfer. In spite of this, there is a wide underlying degree of consensus as to the importance of intellectual property to its possessor as well as for its role in bringing about the desired ends (of technology transfer, sustainable exploitation of genetic resources, etc.). It is assumed that intellectual property rights can fulfil these diverse roles even though the necessary mechanisms remain to be worked out.

Technology transfer, in the most meaningful sense, is much more than the assignment or licensing of intellectual property rights. The transfer of legal rights avails little unless the recipient has the technological capacity to put industrial processes into operation. International conventions are made between governments, but governments alone do not create technology (although the purchase of rights by developed countries and their transfer to developing countries may be a useful element of development co-operation policies). Devising measures to induce private-sector owners of technology to transfer this technology to foreign organisations to the extent envisaged in Article 16 of the Biodiversity Convention will require considerable co-operation among all parties concerned. This is unlikely to be achieved unless all the parties see that real and substantial benefits are to be gained.

The exploitation of genetic resources is an undertaking which, as in all areas of the life sciences, will have its share of failures at the research, development or commercialisation stage. For this reason it is recognised as doubtful that the demands for benefit sharing and technology transfer will alone generate the level of international funding necessary to achieve the primary objective of the Biodiversity Convention. The Global Environment Facility is clearly of central importance in this context.

This report takes a step towards a better understanding of the issues involved and of the positive role of intellectual property rights in the establishment of workable solutions. To make further progress, improved co-ordination among OECD Member countries would be desirable. Such co-ordination was stated at the beginning of the questionnaire as the objective of this OECD activity.

[illegible]

[illegible]

Annex I

STUDY OF INTELLECTUAL PROPERTY POLICY IN THE FIELD OF BIOTECHNOLOGY WITH REGARD TO TECHNOLOGY TRANSFER

QUESTIONNAIRE

Background

Objective of the activity

Inform and co-ordinate the views of OECD Member countries with regard to intellectual property in relation to technology transfer

by conducting a survey of current practices, experiences and expectations related to technology transfer in biotechnology,

which will analyse, from a policy and economic perspective, issues of technology transfer incident to access to genetic resources.

Scope of the activity

The questionnaire relates to laws, agreements, contracts and practices, including those in and between OECD Member countries as well as with other countries, that involve either or both:

- the transfer between countries of technology in the field of biotechnology related to the development of genetic resources;
- access to the genetic resources of one country by persons or organisations based in another country.

Use of terms

For the purposes of this questionnaire, the following terms shall have the following meanings:

"Biotechnology" is the application of scientific and engineering principles to the processing of materials by biological agents to provide goods and services.*

"Genetic material" is any material of plant, animal, microbial or other origin containing functional units of heredity.**

"Genetic resources" is genetic material of actual or potential value.***

"Transfer of technology" includes, but is not limited to, the disclosure of results from research and development, the licensing or assignment of intellectual property rights related to such results, exchange of information, education and training, and joint ventures.

Questions

I. Existing Projects

Please describe the types of projects that you have sponsored, or participated in, related to the discovery or development of genetic resources:

1. Describe the **participants and their roles** (examples of categories of arrangements):

   a) company/company arrangements
   b) university/company arrangements
   c) non-profit organisation/company arrangements
   d) government/company arrangements
   e) government/government arrangements
   f) other ______________________

2. Describe in general terms the arrangements that you have made with regard to **types of technology transfer**:

   a) disclosure of results from research and development
   b) licensing or assignment of intellectual property rights related to such results
   c) exchange of information
   d) education and training
   e) joint ventures

* *Biotechnology – International Trends and Perspectives*, OECD, 1982.

** Convention on Biological Diversity.

*** Convention on Biological Diversity.

f) acquisition of one entity by another
g) financial and other support of research activities
h) other ______________________

3. Describe in general terms the arrangements that you have made with regard to **access to and use of genetic resources:**

   a) consent
   b) compensation and other benefits
   c) ownership and control of genetic material

4. Describe the arrangements that you have made with regard to **ownership, control and protection of intellectual property rights** in agreements relating to access to and use of genetic resources:

   a) patents
   b) protection of undisclosed information
   c) copyright
   d) trade marks or service marks
   e) plant variety protection
   f) other ______________________

II. Evaluation of Government Policies

Please rate the significance of each of the following factors with respect to developing agreements on research or development related to genetic resources.

(A: essential B: major importance C: minor importance D: no importance)

5. Availability of intellectual property protection in the country to which technology is transferred, in the form of:

   A B C D (a) patents
   A B C D (b) protection of undisclosed information
   A B C D (c) copyright
   A B C D (d) trade marks or service marks
   A B C D (e) plant variety protection
   A B C D (f) other ______________________

6. Availability of intellectual property protection in your country, in the form of:

   A B C D (a) patents
   A B C D (b) protection of undisclosed information
   A B C D (c) copyright
   A B C D (d) trade marks or service marks
   A B C D (e) plant variety protection
   A B C D (f) other ______________________

7. Your ability to enforce intellectual property rights effectively:

A B C D	(a)	in the country to which technology is transferred
A B C D	(b)	in your country
A B C D	(c)	in other countries

8. Restrictions placed on marketing products that result from development of genetic resources:

A B C D	(a)	in the country to which technology is transferred
A B C D	(b)	in your country
A B C D	(c)	in other countries

9. Financial and investment conditions, particularly:

A B C D	(a)	availability of grants or subsidies for conducting research or development of genetic resources
A B C D	(b)	tax incentives regarding research or development of genetic resources
A B C D	(c)	other ____________________

10. Conditions placed on parties prior to gaining access to genetic* resources:

A B C D	(a)	requirement of "informed consent" as to possible future use of genetic resources
A B C D	(b)	requirement for disclosure of research results
A B C D	(c)	sharing, either through licenses or partial ownership interest, intellectual property rights to technology developed during the project
A B C D	(d)	other restrictions on the licensing of intellectual property rights
A B C D	(e)	profit-sharing or royalty requirements
A B C D	(f)	up-front or fixed-fee obligations as a condition of access to use of genetic resources
A B C D	(g)	other ____________________

III. Considerations for Future Policies

11. Please describe briefly measures that might be taken:

a) to assist parties concerned with the conservation of genetic resources;
b) to improve the protection of biotechnology world-wide.

* Convention on Biological Diversity.

12. What are the most important challenges facing the legal system for protection of biotechnology:

    a) in the near term?

    b) in the medium term?

Annex II

ARTICLES 15, 16 AND 17 OF THE CONVENTION ON BIOLOGICAL DIVERSITY

The following articles are of relevance to this report.

Article 15: Access to genetic resources

1. Recognizing the sovereign rights of States over their natural resources, the authority to determine access to genetic resources rests with the national governments and is subject to national legislation.

2. Each Contracting Party shall endeavour to create conditions to facilitate access to genetic resources for environmentally sound uses by other Contracting Parties and not to impose restrictions that run counter to the objectives of this Convention.

3. For the purpose of this Convention, the genetic resources being provided by a Contracting Party, as referred to in this Article and Articles 16 and 19, are only those that are provided by Contracting Parties that are countries of origin of such resources or by the Parties that have acquired the genetic resources in accordance with this Convention.

4. Access, where granted, shall be on mutually agreed terms and subject to the provisions of this Article.

5. Access to genetic resources shall be subject to prior informed consent of the Contracting Party providing such resources, unless otherwise determined by that Party.

6. Each Contracting Party shall endeavour to develop and carry out scientific research based on genetic resources provided by other Contracting Parties with the full participation of, and where possible in, such Contracting Parties.

7. Each Contracting Party shall take legislative, administrative or policy measures, as appropriate, and in accordance with Articles 16 and 19 and, where necessary, through the financial mechanism established by Articles 20 and 21 with the aim of sharing in a fair and equitable way the results of research and development and the benefits arising from the commercial and other utiliza-

tion of genetic resources with the Contracting Party providing such resources. Such sharing shall be upon mutually agreed terms.

Article 16: Access to and transfer of technology

1. Each Contracting Party, recognizing that technology includes biotechnology, and that both access to and transfer of technology among Contracting Parties are essential elements for the attainment of the objectives of this Convention, undertakes subject to the provisions of this Article to provide and/or facilitate access for and transfer to other Contracting Parties of technologies that are relevant to the conservation and sustainable use of biological diversity or make use of genetic resources and do not cause significant damage to the environment.

2. Access to and transfer of technology referred to in paragraph 1 above to developing countries shall be provided and/or facilitated under fair and most favourable terms, including on concessional and preferential terms where mutually agreed, and, where necessary, in accordance with the financial mechanism established by Articles 20 and 21. In the case of technology subject to patents and other intellectual property rights, such access and transfer shall be provided on terms which recognize and are consistent with the adequate and effective protection of intellectual property rights. The application of this paragraph shall be consistent with paragraphs 3, 4 and 5 below.

3. Each Contracting Party shall take legislative, administrative or policy measures, as appropriate, with the aim that Contracting Parties, in particular those that are developing countries, which provide genetic resources are provided access to and transfer of technology which makes use of those resources, on mutually agreed terms, including technology protected by patents and other intellectual property rights, where necessary, through the provisions of Articles 20 and 21 and in accordance with international law and consistent with paragraphs 4 and 5 below.

4. Each Contracting Party shall take legislative, administrative or policy measures, as appropriate, with the aim that the private sector facilitates access to, joint development and transfer of technology referred to in paragraph 1 above for the benefit of both governmental institutions and the private sector of developing countries and in this regard shall abide by the obligations included in paragraphs 1, 2 and 3 above.

5. The Contracting Parties, recognizing that patents and other intellectual property rights may have an influence on the implementation of this Convention, shall cooperate in this regard subject to national legislation and international law in order to ensure that such rights are supportive of and do not run counter to its objectives.

Article 17: Exchange of information

1. The Contracting Parties shall facilitate the exchange of information, from all publicly available sources, relevant to the conservation and sustainable use of biological diversity, taking into account the special needs of developing countries.

2. Such exchange of information shall include exchange of results of technical, scientific and socio-economic research, as well as information on training and surveying programmes, specialized knowledge, indigenous and traditional knowledge as such and in combination with the technologies referred to in Article 16, paragraph 1. It shall also, where feasible, include repatriation of information.

Annex III

AGREEMENT ON TRADE-RELATED ASPECTS OF INTELLECTUAL PROPERTY RIGHTS (TRIPS AGREEMENT)

The following articles contained in the TRIPS Agreement are among those which are particularly relevant to this report:

Section 5: Patents

Article 27: Patentable subject matter

1. Subject to the provisions of paragraphs 2 and 3, patents shall be available for any inventions, whether products or processes, in all fields of technology, provided that they are new, involve an inventive step and are capable of industrial application.* Subject to paragraph 4 of Article 65, paragraph 8 of Article 70 and paragraph 3 of this Article, patents shall be available and patent rights enjoyable without discrimination as to the place of invention, the field of technology and whether products are imported or locally produced.

2. Members may exclude from patentability inventions, the prevention within their territory of the commercial exploitation of which is necessary to protect *ordre public* or morality, including to protect human, animal or plant life or health or to avoid serious prejudice to the environment, provided that such exclusion is not made merely because the exploitation is prohibited by their law.

3. Members may also exclude from patentability:
   a) diagnostic, therapeutic and surgical methods for the treatment of humans or animals;
   b) plants and animals other than micro-organisms, and essentially biological processes for the production of plants or animals other than non-

* For the purposes of this Article, the terms "inventive step" and "capable of industrial application" may be deemed by a Member to be synonymous with the terms "non-obvious" and "useful" respectively.

biological and microbiological processes. However, Members shall provide for the protection of plant varieties either by patents or by an effective *sui generis* system or by any combination thereof. The provisions of this subparagraph shall be reviewed four years after the date of entry into force of the WTO Agreement.

Article 33: Term of protection

The term of protection available shall not end before the expiration of a period of twenty years counted from the filing date.*

Article 34: Process patents: burden of proof

1. For the purposes of civil proceedings in respect of the infringement of the rights of the owner referred to in paragraph 1*(b)* of Article 28, if the subject matter of a patent is a process for obtaining a product, the judicial authorities shall have the authority to order the defendant to prove that the process to obtain an identical product is different from the patented process. Therefore, Members shall provide, in at least one of the following circumstances, that any identical product when produced without the consent of the patent owner shall, in the absence of proof to the contrary, be deemed to have been obtained by the patented process:

a) if the product obtained by the patented process is new;

b) if there is a substantial likelihood that the identical product was made by the process and the owner of the patent has been unable through reasonable efforts to determine the process actually used.

2. Any Member shall be free to provide that the burden of proof indicated in paragraph 1 shall be on the alleged infringer only if the condition referred to in subparagraph *(a)* is fulfilled or only if the condition referred to in subparagraph *(b)* is fulfilled.

3. In the adduction of proof to the contrary, the legitimate interests of defendants in protecting their manufacturing and business secrets shall be taken into account.

* It is understood that those Members which do not have a system of original grant may provide that the term of protection shall be computed from the filing date in the system of original grant.

Section 7: Protection of undisclosed information

Article 39

3. Members, when requiring, as a condition of approving the marketing of pharmaceutical or of agricultural chemical products which utilize new chemical entities, the submission of undisclosed test or other data, the origination of which involves a considerable effort, shall protect such data against unfair commercial use. In addition, Members shall protect such data against disclosure, except where necessary to protect the public, or unless steps are taken to ensure that the data are protected against unfair commercial use.

Part VI: Transitional arrangements

Article 65: Transitional arrangements

1. Subject to the provisions of paragraphs 2, 3 and 4, no Member shall be obliged to apply the provisions of this Agreement before the expiry of a general period of one year following the date of entry into force of the WTO Agreement.

2. A developing country Member is entitled to delay for a further period of four years the date of application, as defined in paragraph 1, of the provisions of this Agreement other than Articles 3, 4 and 5.

3. Any other Member which is in the process of transformation from a centrally-planned into a market, free-enterprise economy and which is undertaking structural reform of its intellectual property system and facing special problems in the preparation and implementation of intellectual property laws and regulations, may also benefit from a period of delay as foreseen in paragraph 2.

4. To the extent that a developing country Member is obliged by this Agreement to extend product patent protection to areas of technology not so protectable in its territory on the general date of application of this Agreement for that Member, as defined in paragraph 2, it may delay the application of the provisions on product patents of Section 5 of Part II to such areas of technology for an additional period of five years.

5. A Member availing itself of a transitional period under paragraphs 1, 2, 3 or 4 shall ensure that any changes in its laws, regulations and practice made during that period do not result in a lesser degree of consistency with the provisions of this Agreement.

Article 66: Least-developed country members

1. In view of the special needs and requirements of least-developed country Members, their economic, financial and administrative constraints, and their

need for flexibility to create a viable technological base, such Members shall not be required to apply the provisions of this Agreement, other than Articles 3, 4 and 5, for a period of 10 years from the date of application as defined under paragraph 1 of Article 65. The Council for TRIPS shall, upon duly motivated request by a least-developed country Member, accord extensions of this period.

2. Developed country Members shall provide incentives to enterprises and institutions in their territories for the purpose of promoting and encouraging technology transfer to least-developed country Members in order to enable them to create a sound and viable technological base.

Article 67: Technical cooperation

In order to facilitate the implementation of this Agreement, developed country Members shall provide, on request and on mutually agreed terms and conditions, technical and financial cooperation in favour of developing and least-developed country Members. Such cooperation shall include assistance in the preparation of laws and regulations on the protection and enforcement of intellectual property rights as well as on the prevention of their abuse, and shall include support regarding the establishment or reinforcement of domestic offices and agencies relevant to these matters, including the training of personnel.

Part VII: Institutional arrangements; final provisions

Article 70: Protection of existing subject matter

8. Where a Member does not make available as of the date of entry into force of the WTO Agreement patent protection for pharmaceutical and agricultural chemical products commensurate with its obligations under Article 27, that Member shall:

- *a)* notwithstanding the provisions of Part VI, provide as from the date of entry into force of the WTO Agreement a means by which applications for patents for such inventions can be filed;
- *b)* apply to these applications, as of the date of application of this Agreement, the criteria for patentability as laid down in this Agreement as if those criteria were being applied on the date of filing in that Member or, where priority is available and claimed, the priority date of the application; and
- *c)* provide patent protection in accordance with this Agreement as from the grant of the patent and for the remainder of the patent term, counted from the filing date in accordance with Article 33 of this

Agreement, for those of these applications that meet the criteria for protection referred to in subparagraph *(b)*.

9. Where a product is the subject of a patent application in a Member in accordance with paragraph 8*(a)*, exclusive marketing rights shall be granted, notwithstanding the provisions of Part VI, for a period of five years after obtaining marketing approval in that Member or until a product patent is granted or rejected in that Member, whichever period is shorter, provided that, subsequent to the entry into force of the WTO Agreement, a patent application has been filed and a patent granted for that product in another Member and marketing approval obtained in such other Member.

Annex IV

MATERIAL TRANSFER AGREEMENTS

A type of Material Transfer Agreement and a Standard Order Form under consideration by an International Agricultural Research Centre (referred to below as "the Centre") are given below:

MATERIAL TRANSFER AGREEMENT

The material contained herein is being furnished by (the Centre) under the following conditions:

1. (The Centre) is making the material described in the attached list available as part of its policy of maximising the utilisation of genetic material for research. The material was either developed by (the Centre); or it was acquired prior to the entry into force of the United Nations Convention on Biological Diversity; or if it was acquired after the entering into force of the Biodiversity Convention, it was obtained with the understanding that it could be made freely available for any agricultural research or breeding purposes.

2. The recipient may reproduce the seed and use the material for agricultural research and breeding purposes and may distribute it to other parties provided that any recipient is willing to accept the conditions of this agreement.

3. If the seed packet is labelled "FAO Designated Germplasm" the material is held in trust under the terms of an agreement between (the Centre) and FAO, and the recipient has no rights to obtain intellectual property rights.

4. Recipients are free to release for commercialisation (the Centre) research products in the form they are provided. If released without obtaining Intellectual property rights, (the Centre) requests notification and acknowledgement. Recipients are not to apply for any form of Intellectual property rights of (the Centre) research products without the written permission of (the Centre). Moreover, while (the Centre) recognises the validity of Intellectual property rights, it reserves the right to distribute all material in accordance with paragraph 1 above.

5. (The Centre) makes no warranties as to the safety or title of the material, nor as to the accuracy or correctness of any passport or other data provided with the material. Neither does it make any warranties as to the quality, viability, or purity (genetic or mechanical) of the material being furnished. The phytosanitary condition of the material is warranted only as described in the attached phytosanitary certificate. The recipient assumes full responsibility for complying with the recipient nation's biosafety regulations and rules as to import or release of genetic material.

6. Upon request (the Centre) will furnish information that may be available in addition to whatever is furnished with the seed. Recipients are requested to furnish (the Centre) performance data collected during evaluation.

7. The material is supplied expressly conditional on acceptance of the term of this agreement. The recipient's retention of the material constitutes such acceptance.

Standard Order Form

I/We order the following material:

see attached list

Insofar as this material is "FAO Designated Germplasm" under the Agreement between (the Centre) and the Food and Agriculture Organization of the United Nations (FAO) Placing Collections of Plant Germplasm under the Auspices of FAO dated 26 October 1994, I/We agree

a) not to claim ownership over the material received, nor to seek intellectual property rights over the germplasm or related information,

b) to ensure that any subsequent person or institution to whom I/We make samples of the germplasm available, is bound by the same provision.

Name of person or institution
requesting the germplasm: ______________________

Address: ______________________

Shipping address: ______________________

Authorised signature: ______________________

Date: ______________________

Annex V

LETTER OF COLLECTION

AGREEMENT BETWEEN SOURCE COUNTRY AND DEVELOPMENTAL THERAPEUTICS PROGRAM DIVISION OF CANCER TREATMENT NATIONAL CANCER INSTITUTE, UNITED STATES

1. The Developmental Therapeutics Program (DTP), Division of Cancer Treatment (DCT), National Cancer Institute (NCI) is currently investigating plants, marine macro-organisms and microbes as potential sources of novel anticancer and AIDS-antiviral drugs. The DTP is the drug discovery program of the NCI which is an Institute of the National Institutes of Health (NIH), and arm of the Department of Health and Human Services of the United States Government. While investigating the potential of natural products in drug discovery and development, NCI wishes to promote the conservation of biological diversity, and recognizes the need to compensate source country organizations and peoples in the event of commercialization of a drug developed from an organism collected within their borders.

2. As part of the drug discovery program, DTP has contracts with various organizations for the collection of plants and marine macro-organisms worldwide. DTP has an interest in investigating plants from Source Country, and wishes to collaborate with the Source Country Government ("SCG") or Source Country Organization(s) ("SGO") as appropriate in this investigation. The collection of plants will be within the framework of the collection contract between the NCI and the NCI Contractor (Contractor) which will collaborate with the appropriate agency in the Source Country Government ("SCG") or the Source Country Organization(s) ("SCO"). The NCI will make sincere efforts to transfer knowledge, expertise, and technology related to drug discovery and development to the appropriate Source Country Institution ("SCI") in Source Country as the agent appointed by the Source Country Government ("SCG") or Source Country Organization(s) ("SCO"), subject to the provision of mutually acceptable guarantees for the protection of intellectual property associated with any patented technology. The Source Country Government ("SCG") or Source Country Organization(s) ("SCO"), in turn, desires to collaborate closely with

the DCT/NCI in pursuit of the investigation of its plants, subject to the conditions and stipulations of this agreement.

3. The role of DTP, DCT, NCI in the collaboration will include the following:
   1. DTP/NCI will screen the extracts of all plants provided from Source Country for anticancer and AIDS-antiviral activity, and will provide the test results to SCI on a quarterly basis. Such results will be channelled via Contractor.
   2. The test results will be kept confidential by all parties, with any publication delayed until DTP/NCI has an opportunity to file a patent application in the United States of America on any active agents isolated. Such application will be made according to the terms stated in clause 6.
   3. Any extracts exhibiting significant activity will be further studied by bioassay-guided fractionation in order to isolate the pure compound(s) responsible for the observed activity. Since the relevant bioassays are only available at DTP/NCI, such fractionation will be carried out in DTP/NCI laboratories. A suitably qualified scientist designated by SCI may participate in this process subject to the terms stated in clause 4. In addition, in the course of the contract period, DTP/NCI will assist the Source Country Government ("SCG") or Source Country Organization(s) ("SCO"), in conjunction with SCI, to develop the capacity to undertake drug discovery and development, including capabilities for the screening and isolation of active compounds from plants and marine organisms.
   4. Subject to the provision that suitable laboratory space and other necessary resources are available, DTP/NCI agrees to invite a senior technician or scientist designated by SCI to work in the laboratories of DTP/NCI or, if the parties agree, in laboratories using technology which would be useful in furthering work under this agreement. The duration of such a visit would not exceed one year except by prior agreement between SCI and DTP/NCI. The designated Guest Researcher will be subject to provisions usually governing Guest Researchers at NIH, except when carrying out research on materials provided through collections in Source Country. Salary and other conditions of exchange will be negotiated in good faith.
   5. In the event of the isolation of a promising agent from a plant collected in Source Country, further development of the agent will be undertaken by DTP/NCI in collaboration with SCI. Once an active agent is approved by the DTP/NCI for preclinical development, SCI and the DTP/NCI will discuss participation by SCI scientists in the development of the specific agent.

      The DTP/NCI will make a sincere effort to transfer any knowledge, expertise, and technology developed during such collaboration in the discovery and development process to SCI, subject to the provision of mutually acceptable guarantees for the protection of intellectual property associated with any patented technology.

6. DTP/NCI will, as appropriate, seek patent protection on all inventions developed under this agreement by DTP/NCI employees alone or by DTP/NCI and Source Country Government ("SCG") or Source Country Organization(s) ("SCO") employees jointly, and will seek appropriate protection abroad, including in Source Country, if appropriate.
7. All licenses granted on any patents arising from this collaboration shall contain a clause referring to this agreement and shall indicate that the licensee has been apprised of this agreement.
8. Should the agent eventually be licensed to a pharmaceutical company for production and marketing, DTP/NCI will require the successful licensee to negotiate and enter into agreement(s) with the Source Country Government ("SCG") agency(ies) or Source Country Organization(s) ("SCO") as appropriate. This agreement(s) will address the concern on the part of the Source Country Government ("SCG") or Source Country Organization(s) ("SCO"), that pertinent agencies, institutions and/or persons receive royalties and other forms of compensation, as appropriate.
9. Such terms shall apply equally to instances where the invention is the actual isolated natural product, or, where the invention is a product structurally based on the isolated natural product (*i.e.* where the natural product provides the lead for the development of invention), though the percentage of royalties negotiated as payment might vary depending upon the relationship of the marketed drug to the originally isolated product. It is understood that the eventual development of a drug to the stage of marketing is a long-term process which may require 10-15 years.
10. In obtaining licensees, the DTP/NCI will require the applicant for license to seek as its first source of supply the natural products from Source Country. If no appropriate licensee is found who will use natural products available from Source Country, of if the Source Country Government ("SCG") or Source Country Organization(s) ("SCO") as appropriate, or its suppliers cannot provide adequate amounts of raw materials at a mutually agreeable fair price, the licensee will be required to pay to Source Country Government ("SCG") or Source Country Organization(s) ("SCO") as appropriate, an amount of money (to be negotiated) to be used for expenses associated with cultivation of medicinal plant species that are endangered by deforestation, or for other appropriate conservation measures. Such terms will also apply to instances where the active agent is prepared by total synthesis.
11. Section 10 shall not apply to organisms which are freely available from different countries (*i.e.* common weeds, agricultural crops, ornamental plants, fouling organisms) unless information indicating a particular use of the organism (*e.g.* medicinal, pesticidal) was provided by local residents to guide the collection of such an organism from Source Country, or unless

other justification acceptable to both the Source Country Government ("SCG") or Source Country Organization(s) ("SCO") and the DTP/NCI is provided. In the case where an organism is freely available from different countries, but a genotype producing an active agent is found only in the Source Country, Section 10 shall apply.

12. DTP/NCI will test any pure compounds submitted by the Source Country Government ("SCG") or Source Country Organization(s) ("SCO") as appropriate and SCI scientists for antitumor and AIDS-antiviral activity, provided such compounds have not been tested previously in the DTP/NCI screens. If significant antitumor or AIDS-antiviral activity is detected, further development of the compound and investigation of patent rights will, as appropriate, be undertaken by DTP/NCI in consultation with SCI and the Source Country Government ("SCG") or Source Country Organization(s) ("SCO").
13. Should the agent eventually be licensed to a pharmaceutical company for production and marketing, DTP/NCI will require the successful licensee to negotiate and enter into agreement(s) with the appropriate Source Country Government ("SCG") agency(ies) or Source Country Organization(s) ("SCO"). This agreement will address the concern on the part of the Source Country Government ("SCG") or Source Country Organization(s) ("SCO") that pertinent agencies, institutions and/or persons receive royalties and other forms of compensation, as appropriate.
14. DTP/NCI may send selected samples to other organizations for investigation of their anticancer, anti-HIV or other therapeutic potential. Such samples will be restricted to those collected by NCI contractors unless specifically authorized by the Source Country Government ("SGO") or Source Country Organization(s) ("SCO"). Any organization receiving samples must agree to compensate the Source Country Government ("SCG") or Source Country Organization(s) ("SCO") and individuals, as appropriate, in the same fashion as described in Sections 8-10 above, notwithstanding anything to the contrary in Section 11.

The role of the Source Country Government ("SCG") or Source Country Organization(s) ("SCO") in the collaboration will include the following:

1. The appropriate agency in Source Country Government ("SCG") or the Source Country Organization(s) ("SCO") will collaborate with Contractor in the collection of plants, and will work with Contractor to arrange the necessary permits to ensure the timely collection and export of materials to DTP/NCI.
2. Should the appropriate agency in Source Country Government ("SCG") or the Source Country Organization(s) ("SCO") have any knowledge of the medicinal use of any plants by the local population or traditional healers, this information will be used to guide the collection of plants on a priority basis where possible. Details of the methods of adminis-

tration (*e.g.* hot fusion, etc.) used by the traditional healers will be provided where applicable to enable suitable extracts to be made. All such information will be kept confidential by DTP/NCI until both parties agree to publication.

The permission of the traditional healer or community will be sought before publication of their information, and proper acknowledgement will be made of their contribution.

3. The appropriate agency in Source Country Government ("SCG") or Source Country Organization(s) ("SCO") and Contractor will collaborate in the provision of further quantities of active raw material if required for development studies.
4. In the event of large amounts of raw material being required for production, the appropriate agency in Source Country Government ("SCG") or Source Country Organization(s) ("SCO") and Contractor will investigate the mass propagation of the material in Source Country. Consideration should also be given to sustainable harvest of the material while conserving the biological diversity of the region, and involvement of the local population in the planning and implementation stages.
5. Source Country Government ("SCG") or Source Country Organization(s) ("SCO") and SCI scientists and their collaborators may screen additional samples of the same raw materials for other biological activities and develop them for such purposes independently of this agreement.

This agreement may be amended at any time subject to the written agreement of both parties.

	________________ Name (Signature)
	________________ Name (Print or type)
	________________ Institution or Agency
	________________ Institution or Agency
________________ Director, National Cancer Institute	________________ Address
________________ Date	________________ Date

Annex VI

INTERNATIONAL UNION FOR THE PROTECTION OF NEW VARIETIES OF PLANT (UPOV), 1991

THE RIGHTS OF THE BREEDER

The following articles contained in the 1991 Revision of the UPOV International Convention are among those which are particularly relevant to this report. This revision requires ratification by UPOV Member States.

Article 14: Scope of the breeder's right

1. [Acts in respect of the propagating material]

   a) Subject to Articles 15 and 16, the following acts in respect of the propagating material of the protected variety shall require the authorization of the breeder:

   i) production or reproduction (multiplication),
   ii) conditioning for the purpose of propagation,
   iii) offering for sale,
   iv) selling or other marketing,
   v) exporting,
   vi) importing,
   vii) stocking for any of the purposes mentioned in *(i)* to *(vi)* above.

   b) The breeder may make his authorization subject to conditions and limitations.

2. [Acts in respect of the harvested material] Subject to Articles 15 and 16, the acts referred to in items *(i)* to *(vii)* of paragraph (1)*(a)* in respect of harvested material, including entire plants and parts of plants, obtained through the unauthorized use of propagating material of the protected variety shall require the authorization of the breeder, unless the breeder has had reasonable opportunity to exercise his right in relation to the said propagating material.

3. [Acts in respect of certain products] Each Contracting Party may provide that, subject to Articles 15 and 16, the acts referred to in items *(i)* to *(vii)* of paragraph (1)*(a)* in respect of products made directly from harvested material

of the protected variety falling within the provisions of paragraph (2) through the unauthorized use of the said harvested material shall require the authorization of the breeder unless the breeder has had reasonable opportunity to exercise his right in relation to the said harvested material.

4. [Possible additional acts] Each Contracting Party may provide that, subject to Articles 15 and 16, acts other than those referred to in items *(i)* to *(vii)* of paragraph (1)*(a)* shall also require the authorization of the breeder.

5. [Essentially derived and certain other varieties]

a) The provisions of paragraphs (1) to (4) shall also apply in relation to:

i) varieties which are essentially derived from the protected variety, where the protected variety is not itself an essentially derived variety,

ii) varieties which are not clearly distinguishable in accordance with Article 7 from the protected variety and

iii) varieties whose production requires the repeated use of the protected variety.

b) For the purposes of sub-paragraph *(a)(i)*, a variety shall be deemed to be essentially derived from another variety ("the initial variety") when

i) it is predominantly derived from the initial variety, or from a variety that is itself predominantly derived from the initial variety, while retaining the expression of the essential characteristics that result from the genotype or combination of genotypes of the initial variety,

ii) it is clearly distinguishable from the initial variety and

iii) except for the differences which result from the act of derivation, it conforms to the initial variety in the expression of the essential characteristics that result from the genotype or combination of genotypes of the initial variety.

c) Essentially derived varieties may be obtained for example by the selection of a natural or induced mutant, or of a somaclonal variant, the selection of a variant individual from plants of the initial variety, backcrossing, or transformation by genetic engineering.

Article 15: Exceptions to the breeder's right

1. [Compulsory exceptions] The breeder's right shall not extend to

i) acts done privately and for non-commercial purposes,

ii) acts done for experimental purposes and

iii) acts done for the purpose of breeding other varieties, and, except where the provisions of Article 14(5) apply, acts referred to in Article 14(1) to (4) in respect of such other varieties.

2. [Optional exception] Notwithstanding Article 14, each Contracting Party may, within reasonable limits and subject to the safeguarding of the legitimate interests of the breeder, restrict the breeder's right in relation to any variety in order to permit farmers to use for propagating purposes, on their own holdings, the product of the harvest which they have obtained by planting, on their own holdings, the protected variety or a variety covered by Article 14(5)*(a)(i)* or *(ii)*.

Article 16: Exhaustion of the breeder's right

1. [Exhaustion of right] The breeder's right shall not extend to acts concerning any material of the protected variety, or of a variety covered by the provisions of Article 14(5), which has been sold or otherwise marketed by the breeder or with his consent in the territory of the Contracting Party concerned, or any material derived from the said material, unless such acts

i) involve further propagation of the variety in question or

ii) involve an export of material of the variety, which enables the propagation of the variety, into a country which does not protect varieties of the plant genus or species to which the variety belongs, except where the exported material is for final consumption purposes.

Libreria Hoepli
Via Hoepli 5
20121 Milano Tel. (02) 86.54.46
Fax: (02) 805.28.86

Libreria Scientifica
Dott. Lucio de Biasio 'Aeiou'
Via Coronelli, 6
20146 Milano Tel. (02) 48.95.45.52
Fax: (02) 48.95.45.48

JAPAN – JAPON
OECD Tokyo Centre
Landic Akasaka Building
2-3-4 Akasaka, Minato-ku
Tokyo 107 Tel. (81.3) 3586.2016
Fax: (81.3) 3584.7929

KOREA – CORÉE
Kyobo Book Centre Co. Ltd.
P.O. Box 1658, Kwang Hwa Moon
Seoul Tel. 730.78.91
Fax: 735.00.30

MALAYSIA – MALAISIE
University of Malaya Bookshop
University of Malaya
P.O. Box 1127, Jalan Pantai Baru
59700 Kuala Lumpur
Malaysia Tel. 756.5000/756.5425
Fax: 756.3246

MEXICO – MEXIQUE
OECD Mexico Centre
Edificio INFOTEC
Av. San Fernando no. 37
Col. Toriello Guerra
Tlalpan C.P. 14050
Mexico D.F. Tel. (525) 665 47 99
Fax: (525) 606 13 07

NETHERLANDS – PAYS-BAS
SDU Uitgeverij Plantijnstraat
Externe Fondsen
Postbus 20014
2500 EA's-Gravenhage Tel. (070) 37.89.880
Voor bestellingen: Fax: (070) 34.75.778

Subscription Agency/Agence d'abonnements :
SWETS & ZEITLINGER BV
Heereweg 347B
P.O. Box 830
2160 SZ Lisse Tel. 252.435.111
Fax: 252.415.888

NEW ZEALAND – NOUVELLE-ZÉLANDE
GPLegislation Services
P.O. Box 12418
Thorndon, Wellington Tel. (04) 496.5655
Fax: (04) 496.5698

NORWAY – NORVÈGE
NIC INFO A/S
Ostensjoveien 18
P.O. Box 6512 Etterstad
0606 Oslo Tel. (22) 97.45.00
Fax: (22) 97.45.45

PAKISTAN
Mirza Book Agency
65 Shahrah Quaid-E-Azam
Lahore 54000 Tel. (42) 735.36.01
Fax: (42) 576.37.14

PHILIPPINE – PHILIPPINES
International Booksource Center Inc.
Rm 179/920 Cityland 10 Condo Tower 2
HV dela Costa Ext cor Valero St.
Makati Metro Manila Tel. (632) 817 9676
Fax: (632) 817 1741

POLAND – POLOGNE
Ars Polona
00-950 Warszawa
Krakowskie Prezdmiescie 7 Tel. (22) 264760
Fax: (22) 265334

PORTUGAL
Livraria Portugal
Rua do Carmo 70-74
Apart. 2681
1200 Lisboa Tel. (01) 347.49.82/5
Fax: (01) 347.02.64

SINGAPORE – SINGAPOUR
Ashgate Publishing
Asia Pacific Pte. Ltd
Golden Wheel Building, 04-03
41, Kallang Pudding Road
Singapore 349316 Tel. 741.5166
Fax: 742.9356

SPAIN – ESPAGNE
Mundi-Prensa Libros S.A.
Castelló 37, Apartado 1223
Madrid 28001 Tel. (91) 431.33.99
Fax: (91) 575.39.98

Mundi-Prensa Barcelona
Consell de Cent No. 391
08009 – Barcelona Tel. (93) 488.34.92
Fax: (93) 487.76.59

Llibreria de la Generalitat
Palau Moja
Rambla dels Estudis, 118
08002 – Barcelona
(Subscripcions) Tel. (93) 318.80.12
(Publicacions) Tel. (93) 302.67.23
Fax: (93) 412.18.54

SRI LANKA
Centre for Policy Research
c/o Colombo Agencies Ltd.
No. 300-304, Galle Road
Colombo 3 Tel. (1) 574240, 573551-2
Fax: (1) 575394, 510711

SWEDEN – SUÈDE
CE Fritzes AB
S–106 47 Stockholm Tel. (08) 690.90.90
Fax: (08) 20.50.21

For electronic publications only/
Publications électroniques seulement
STATISTICS SWEDEN
Informationsservice
S-115 81 Stockholm Tel. 8 783 5066
Fax: 8 783 4045

Subscription Agency/Agence d'abonnements :
Wennergren-Williams Info AB
P.O. Box 1305
171 25 Solna Tel. (08) 705.97.50
Fax: (08) 27.00.71

SWITZERLAND – SUISSE
Maditec S.A. (Books and Periodicals/Livres et périodiques)
Chemin des Palettes 4
Case postale 266
1020 Renens VD 1 Tel. (021) 635.08.65
Fax: (021) 635.07.80

Librairie Payot S.A.
4, place Pépinet
CP 3212
1002 Lausanne Tel. (021) 320.25.11
Fax: (021) 320.25.14

Librairie Unilivres
6, rue de Candolle
1205 Genève Tel. (022) 320.26.23
Fax: (022) 329.73.18

Subscription Agency/Agence d'abonnements :
Dynapresse Marketing S.A.
38, avenue Vibert
1227 Carouge Tel. (022) 308.08.70
Fax: (022) 308.07.99

See also – Voir aussi :
OECD Bonn Centre
August-Bebel-Allee 6
D-53175 Bonn (Germany)
Tel. (0228) 959.120
Fax: (0228) 959.12.17

THAILAND – THAÏLANDE
Suksit Siam Co. Ltd.
113, 115 Fuang Nakhon Rd.
Opp. Wat Rajbopith
Bangkok 10200 Tel. (662) 225.9531/2
Fax: (662) 222.5188

TRINIDAD & TOBAGO, CARIBBEAN
TRINITÉ-ET-TOBAGO, CARAÏBES
SSL Systematics Studies Limited
9 Watts Street
Curepe, Trinadad & Tobago, W.I.
Tel. (1809) 645.3475
Fax: (1809) 662.5654

TUNISIA – TUNISIE
Grande Librairie Spécialisée
Fendri Ali
Avenue Haffouz Imm El-Intilaka
Bloc B 1 Sfax 3000 Tel. (216-4) 296 855
Fax: (216-4) 298.270

TURKEY – TURQUIE
Kültür Yayinlari Is-Türk Ltd. Sti.
Atatürk Bulvari No. 191/Kat 13
06684 Kavaklidere/Ankara
Tél. (312) 428.11.40 Ext. 2458
Fax : (312) 417.24.90
et 425.07.50-51-52-53

Dolmabahce Cad. No. 29
Besiktas/Istanbul Tel. (212) 260 7188

UNITED KINGDOM – ROYAUME-UNI
HMSO
Gen. enquiries Tel. (0171) 873 0011
Fax: (0171) 873 8463

Postal orders only:
P.O. Box 276, London SW8 5DT
Personal Callers HMSO Bookshop
49 High Holborn, London WC1V 6HB

Branches at: Belfast, Birmingham, Bristol, Edinburgh, Manchester

UNITED STATES – ÉTATS-UNIS
OECD Washington Center
2001 L Street N.W., Suite 650
Washington, D.C. 20036-4922
Tel. (202) 785.6323
Fax: (202) 785.0350

Internet: washcont@oecd.org
Subscriptions to OECD periodicals may also be placed through main subscription agencies.

Les abonnements aux publications périodiques de l'OCDE peuvent être souscrits auprès des principales agences d'abonnement.

Orders and inquiries from countries where Distributors have not yet been appointed should be sent to: OECD Publications, 2, rue André-Pascal, 75775 Paris Cedex 16, France.

Les commandes provenant de pays où l'OCDE n'a pas encore désigné de distributeur peuvent être adressées aux Éditions de l'OCDE, 2, rue André-Pascal, 75775 Paris Cedex 16, France.

8-1996

OECD PUBLICATIONS, 2, rue André-Pascal, 75775 PARIS CEDEX 16
PRINTED IN FRANCE
(93 96 05 1) ISBN 92-64-15328-4 – No. 49055 1996
ISSN 0000-0000

MAIN SALES OUTLETS OF OECD PUBLICATIONS
PRINCIPAUX POINTS DE VENTE DES PUBLICATIONS DE L'OCDE

AUSTRALIA – AUSTRALIE
D.A. Information Services
648 Whitehorse Road, P.O.B 163
Mitcham, Victoria 3132 Tel. (03) 9210.7777
Fax: (03) 9210.7788

AUSTRIA – AUTRICHE
Gerold & Co.
Graben 31
Wien I Tel. (0222) 533.50.14
Fax: (0222) 512.47.31.29

BELGIUM – BELGIQUE
Jean De Lannoy
Avenue du Roi, Koningslaan 202
B-1060 Bruxelles
Tel. (02) 538.51.69/538.08.41
Fax: (02) 538.08.41

CANADA
Renouf Publishing Company Ltd.
1294 Algoma Road
Ottawa, ON K1B 3W8 Tel. (613) 741.4333
Fax: (613) 741.5439

Stores:
61 Sparks Street
Ottawa, ON K1P 5R1 Tel. (613) 238.8985

12 Adelaide Street West
Toronto, ON M5H 1L6 Tel. (416) 363.3171
Fax: (416)363.59.63

Les Éditions La Liberté Inc.
3020 Chemin Sainte-Foy
Sainte-Foy, PQ G1X 3V6 Tel. (418) 658.3763
Fax: (418) 658.3763

Federal Publications Inc.
165 University Avenue, Suite 701
Toronto, ON M5H 3B8 Tel. (416) 860.1611
Fax: (416) 860.1608

Les Publications Fédérales
1185 Université
Montréal, QC H3B 3A7 Tel. (514) 954.1633
Fax: (514) 954.1635

CHINA – CHINE
China National Publications Import
Export Corporation (CNPIEC)
16 Gongti E. Road, Chaoyang District
P.O. Box 88 or 50
Beijing 100704 PR Tel. (01) 506.6688
Fax: (01) 506.3101

CHINESE TAIPEI – TAIPEI CHINOIS
Good Faith Worldwide Int'l. Co. Ltd.
9th Floor, No. 118, Sec. 2
Chung Hsiao E. Road
Taipei Tel. (02) 391.7396/391.7397
Fax: (02) 394.9176

CZECH REPUBLIC – RÉPUBLIQUE TCHÈQUE
National Information Centre
NIS – prodejna
Konviktská 5
Praha 1 – 113 57 Tel. (02) 24.23.09.07
Fax: (02) 24.22.94.33
(*Contact* Ms Jana Pospisilova, nkposp@dec.niz.cz)

DENMARK – DANEMARK
Munksgaard Book and Subscription Service
35, Nørre Søgade, P.O. Box 2148
DK-1016 København K Tel. (33) 12.85.70
Fax: (33) 12.93.87

J. H. Schultz Information A/S,
Herstedvang 12,
DK – 2620 Albertslung Tel. 43 63 23 00
Fax: 43 63 19 69
Internet: s-info@inet.uni-c.dk

EGYPT – ÉGYPTE
The Middle East Observer
41 Sherif Street
Cairo Tel. 392.6919
Fax: 360-6804

FINLAND – FINLANDE
Akateeminen Kirjakauppa
Keskuskatu 1, P.O. Box 128
00100 Helsinki

Subscription Services/Agence d'abonnements :
P.O. Box 23
00371 Helsinki Tel. (358 0) 121 4416
Fax: (358 0) 121.4450

FRANCE
OECD/OCDE
Mail Orders/Commandes par correspondance :
2, rue André-Pascal
75775 Paris Cedex 16 Tel. (33-1) 45.24.82.00
Fax: (33-1) 49.10.42.76
Telex: 640048 OCDE
Internet: Compte.PUBSINQ@oecd.org

Orders via Minitel, France only/
Commandes par Minitel, France exclusivement :
36 15 OCDE

OECD Bookshop/Librairie de l'OCDE :
33, rue Octave-Feuillet
75016 Paris Tél. (33-1) 45.24.81.81
(33-1) 45.24.81.67

Dawson
B.P. 40
91121 Palaiseau Cedex Tel. 69.10.47.00
Fax: 64.54.83.26

Documentation Française
29, quai Voltaire
75007 Paris Tel. 40.15.70.00

Economica
49, rue Héricart
75015 Paris Tel. 45.75.05.67
Fax: 40.58.15.70

Gibert Jeune (Droit-Économie)
6, place Saint-Michel
75006 Paris Tel. 43.25.91.19

Librairie du Commerce International
10, avenue d'Iéna
75016 Paris Tel. 40.73.34.60

Librairie Dunod
Université Paris-Dauphine
Place du Maréchal-de-Lattre-de-Tassigny
75016 Paris Tel. 44.05.40.13

Librairie Lavoisier
11, rue Lavoisier
75008 Paris Tel. 42.65.39.95

Librairie des Sciences Politiques
30, rue Saint-Guillaume
75007 Paris Tel. 45.48.36.02

P.U.F.
49, boulevard Saint-Michel
75005 Paris Tel. 43.25.83.40

Librairie de l'Université
12*a*, rue Nazareth
13100 Aix-en-Provence Tel. (16) 42.26.18.08

Documentation Française
165, rue Garibaldi
69003 Lyon Tel. (16) 78.63.32.23

Librairie Decitre
29, place Bellecour
69002 Lyon Tel. (16) 72.40.54.54

Librairie Sauramps
Le Triangle
34967 Montpellier Cedex 2
Tel. (16) 67.58.85.15
Fax: (16) 67.58.27.36

A la Sorbonne Actual
23, rue de l'Hôtel-des-Postes
06000 Nice Tel. (16) 93.13.77.75
Fax: (16) 93.80.75.69

GERMANY – ALLEMAGNE
OECD Bonn Centre
August-Bebel-Allee 6
D-53175 Bonn Tel. (0228) 959.120
Fax: (0228) 959.12.17

GREECE – GRÈCE
Librairie Kauffmann
Stadiou 28
10564 Athens Tel. (01) 32.55.321
Fax: (01) 32.30.320

HONG-KONG
Swindon Book Co. Ltd.
Astoria Bldg. 3F
34 Ashley Road, Tsimshatsui
Kowloon, Hong Kong Tel. 2376.2062
Fax: 2376.0685

HUNGARY – HONGRIE
Euro Info Service
Margitsziget, Európa Ház
1138 Budapest Tel. (1) 111.62.16
Fax: (1) 111.60.61

ICELAND – ISLANDE
Mál Mog Menning
Laugavegi 18, Pósthólf 392
121 Reykjavik Tel. (1) 552.4240
Fax: (1) 562.3523

INDIA – INDE
Oxford Book and Stationery Co.
Scindia House
New Delhi 110001 Tel. (11) 331.5896/5308
Fax: (11) 371.8275

17 Park Street
Calcutta 700016 Tel. 240832

INDONESIA – INDONÉSIE
Pdii-Lipi
P.O. Box 4298
Jakarta 12042 Tel. (21) 573.34.67
Fax: (21) 573.34.67

IRELAND – IRLANDE
Government Supplies Agency
Publications Section
4/5 Harcourt Road
Dublin 2 Tel. 661.31.11
Fax: 475.27.60

ISRAEL – ISRAËL
Praedicta
5 Shatner Street
P.O. Box 34030
Jerusalem 91430 Tel. (2) 52.84.90/1/2
Fax: (2) 52.84.93

R.O.Y. International
P.O. Box 13056
Tel Aviv 61130 Tel. (3) 546 1423
Fax: (3) 546 1442

Palestinian Authority/Middle East:
INDEX Information Services
P.O.B. 19502
Jerusalem Tel. (2) 27.12.19
Fax: (2) 27.16.34

ITALY – ITALIE
Libreria Commissionaria Sansoni
Via Duca di Calabria 1/1
50125 Firenze Tel. (055) 64.54.15
Fax: (055) 64.12.57

Via Bartolini 29
20155 Milano Tel. (02) 36.50.83

Editrice e Libreria Herder
Piazza Montecitorio 120
00186 Roma Tel. 679.46.28
Fax: 678.47.51

Snow and Ash

An Endless Winter Novel

By Theresa Shaver

Cover art by Melchelle Designs

Author's Note

Hello!

It's been a year since my last book was published and I'm happy to be back!

If you've read any of the Stranded novels then you know I like to start a book off with a little chat. I like to give a brief overview of what I was thinking when I wrote the book and why I went in certain directions.

Let's start with the most important part, THANK YOU! Thank you for reading my stories and for all the feedback and comments you give. You probably have no idea just how important that is to me. I'm not a professional, educated writer. I'm a stay at home Mom with a big imagination. I read every single review and comment made and then go back and reread them when I start a new book. I take the advice and constructive criticisms from the most important people, you the reader, and try to apply them to my writing. So thanks, teacher, I'm getting better because of you! Another reason reviews are so important is that authors need them on their books to be accepted by book marketing websites. Self-published authors like myself are responsible for all the marketing which means even when paying for all the ads, they won't run them unless there are reviews on the books. So again, thank you for that help. Even a one sentence review makes a big difference.

So, this book! Wow, what a difference from the Stranded books. It's a different style altogether and timelines are not aligned in the first half…sorry but it shouldn't be confusing because it's from two different, clearly marked, character's POVs. The first half of the story builds the characters and if you think it's too slow…wait for it! The story aligns and we're off to the races lol.

The other major feedback I got from beta readers makes me nod my head in total agreement. I write detail "lite". Yup, sure do and always will. I've never enjoyed a book that is over the top in details. I don't care what the room looks like, tell me what's going to happen in the room! So that's how I write.

Most importantly, this is book one so the characters are just getting started on their journey. I hope you will stay with them and me as they grow and change over the course of the series.

Book 2 will be out ASAP or at least less than a year. I'm in it right now and it's smooth sailing so far so who knows could be a really quick turnaround.

You guys ROCK!

Theresa

**Book Two, Rain And Ruin, of the Endless Winter Series, is available now.*

Part One

Chapter One-Skylar

Sometimes I think that all the colours in the world died with my mother. She was an amazing artist that painted vivid natural landscapes. The colours of her summer meadow paintings were so lush you could almost feel the softness of the flowers and dampness of the dew on the leaves. I haven't seen such beauty in the real world in seven years.

I was ten when the bombs dropped and man destroyed the world, changing all of the earth's glory and seasons into one long grey winter. I still don't really know what happened to make someone push that fateful button and start the chain reaction that would send hundreds of nuclear bombs sailing through the skies to end everything I've ever known. I was a child and all I was worried about was my next dance class and if I would be invited to Sara Dresden's sleepover. I never knew that the sun could be taken away and that everything could die. I haven't been a child since that day.

My parents were polar opposites. My mother was an artist, always seeing the beauty in the world. My father, a former soldier that saw the ugliness. She used to say that they were the perfect balance. Her light lit his darkness and his dark kept her on earth instead of living in the sky with the sun.

I had no idea what was coming but my father did and he prepared for it, which is why I'm alive today when millions have died.

It was late August and my mom and I were headed to the mall for back to school clothes and supplies. She was only three weeks away from her due date and we wanted to get everything done before my new little brother made his grand entrance. I was excited about having a new brother. I had been an only child for ten years and thought it would be great to play the role of big sister. We had just stepped out of the house and were headed to the car when we heard the phone ringing through the open window. My Dad was in the house so we kept going thinking he would answer it. I had just opened the car door when the front door banged open and my Dad let out a shout, stopping us in our tracks.

"Van! I need you and Sky back inside right now!"

My mom shot me a look over the roof of the car but I just shrugged my shoulders and we closed the doors of the car and headed back into the house. Dad had disappeared but we could hear him banging around in the basement. I remember being impatient because I wanted to get to the mall. That changed to worry when he bounded up the stairs carrying two huge duffle bags. I knew what those bags were. He had called them bug out bags and gave me strict orders to never open them.

My mom's forehead was crinkled with concern and all she got out was "Sweetie, what…?" before he started to bark orders out at us.

"You two have exactly five minutes to get anything you don't want to live without for the rest of your lives packed up. We have to go…NOW!" He yelled when we just stared at him. My Mom put her arm around my shoulders and gave him a sharp look.

"Daniel, explain what's going on. You're scaring Skylar!"

I wasn't really scared but I was edging that way when I saw the flash of fear pass through my dad's eyes. He took a deep breath before explaining why our world was about to end.

"I'm sorry Vanessa, Sky. Things are…I got a call from Bill. He said…"

I stopped edging towards scared and dove in head first. Dad was always calm, always prepared and he never looked scared. Uncle Bill was an old friend of his and my godfather. He was still in the military so whatever he had told my dad for him to react this way must be pretty bad. I felt Mom's hand tighten on my shoulder as we watched Dad struggle to find the words to explain. He finally just shook his head and blurted it out.

"The first nuclear bomb dropped ten minutes ago and there are more in the air. It won't be long before some start heading to this continent. We have to go. We have to get out now before that happens. I need you guys to pack what you can in less than five minutes!"

Mom's grip on my arm was almost painful when she turned it into a soft shove towards the stairs. All she said was "Go", and we were scrambling up the stairs to our rooms. I heard the front door slam open as Dad ran out to his old restored pickup truck to dump the bug out bags. I spun in place in my room not knowing where to start. His yell of "Four minutes!" had me diving for my backpack that was supposed to be replaced at the mall today. I started stuffing my favorite outfits into it as my eyes scanned for the treasures I couldn't live without. There were too many things I wanted but didn't have room for so I just started to grab stuff through a glaze of tears. I didn't really understand then that I would never see my room again. I was just responding to the panic in my parent's voices. I dashed to the door but stopped and turned back for one last look. My eyes settled on a shelf above my bed that held my greatest childhood treasure. His soft blurry orange eyes stared back at me. Mr. Quackers had been retired only a year ago after sleeping next to me since I came home from the hospital as a newborn. His plush yellow fur was rubbed shiny smooth in some places and his beak was threadbare at the tip from being gummed by me as a toddler. I was across the room in a heartbeat and he found himself thrust into my pack before I ran out of the room.

Mom was dragging a suitcase down the hall and her big belly led the way past the open door of the new nursery.

"Mom, what about the baby stuff?!"

She froze in place and I'll never forget the haunted look in her eyes as she turned her head to look into the freshly decorated room that would never be used. After a short pause, she just shook her head and headed down the stairs. I watched her waddle away for a second before glancing back at the nursery door. It wasn't fair! She had put so much time and effort into painting murals on the walls and now my little brother wouldn't get to grow up in it. I hesitated for a split second when Dad yelled out the two-minute warning and then barreled into the room. There were reusable shopping bags against the wall still filled with baby clothes she hadn't put away yet, so I grabbed three of them and started to stuff them with the diapers that had been stacked under the change table. Anything within reach went into the bags from pacifiers to a stuffed monkey and baby ointments. I was tying the overstuffed bags closed with the handles when my Dad started to bellow my name. Hooking them over my arms I shouldered my backpack and flew out of the room and down the stairs. He waved me past him and out the front door before slamming it closed and reaching for the bags hanging from my arms. Everything was thrown into the truck bed before we jumped into the cab. One of the things he did made me realize that this was really serious and we might never come home again. Dad didn't lock the front door of the house.

Driving through the city made me question everything Dad had said. There were no people driving like mad or signs of panic. It was just business as usual. I watched out my window as people did regular people things like line up in drive-thrus and push strollers down the sidewalks. My parents weren't talking and after we cleared the last suburb and the mountains could be seen in the distance, I finally broke the silence.

"Dad, what if Uncle Bill was wrong? It doesn't seem like anyone else is bugging out like us."

I could see his knuckles turn white on the steering wheel before he answered me in a controlled voice.

"No one will know until it's too late. We're lucky he got the call out to us when he did. Trust me, Skylar, that's not something he would be wrong about."

I sat back and bit my lip in confusion.

"Where are we going then?"

It was Mom who turned around and answered.

"We're going to the Man Cave."

My eyebrows shot up in surprise. According to Mom, most men had a man cave in their basements or garages. My dad went and actually bought a cave in the mountains. Dad loved to hunt, fish and camp but wasn't blessed with women folk who shared this passion. Mom and I enjoyed going to museums, flea markets and movies so we had never been to his man cave. I had heard them talking about it sometimes but the only thing I could picture was a dark damp cave that bats flew in and out of. This was beginning to sound like a practical joke to me. Were we really going to go hide in a cave as bombs dropped around us and Mom gave birth to a baby on a stone floor? This had to be a joke, right?

Nothing was funny about the cars ahead of us suddenly going out of control. My mouth gaped open in shock as a minivan swerved off the road and flipped as it hit the ditch. My Dad was yelling for us to hold on as he braked hard and pulled to the side of the road. I didn't realize that I was rocking back and forth in my seat or whimpering as a semi-truck transport rocked past us and jack-knifed to the side. It was like slow motion as the long trailer tipped over and slid down the hi-way sideways brushing all the vehicles ahead of it like a huge broom.

We sat in our truck just staring at the devastation ahead of us for a few minutes until my Mom broke the silence.

"What…what just happened?" It came out a whisper but sounded like a shout to me in the dead silence and I think I flinched.

Dad had his head in his hands and he scrubbed at his face before turning the engine over and putting the truck into drive. He didn't look at Mom as the truck crept ahead towards the wreckage.

"EMP. It's what happens when nuclear bombs drop. They send out a pulse that fries anything electronic. Every car and truck just died as they were going a hundred kilometers an hour."

Mom looked at him in disbelief.

"Then why is our truck still working?"

My eyes were glued to the crashed van we were passing but I heard him quietly say, "Because I built it to still work."

Dad was slowly passing the upside-down van when I saw movement in the crash. A lady was wiggling out of a shattered window and I screamed at him to stop.

"Dad, STOP! We have to help them!"

Mom's face whipped towards the crash we were passing and then to Dad's face. I couldn't see him but whatever she saw there made her take a shuddering breath and look out the opposite window and away from the crashed van. My mouth dropped open in confused shock as he kept driving.

"What? What are you doing? Why aren't you stopping? We have to help those people!"

He didn't answer me but his knuckles flared white against the wheel again. I sat back and let the tears flow down my face and kept my eyes glued to my lap. I couldn't bear to see the suffering we were driving past. I didn't understand then how much in the world had changed when the bombs fell. I still believed in the good of the world, of helping your fellow man in need. It wasn't until much later that I learned that to survive you could only help yourself and the people you loved. At ten years old I couldn't grasp the scale of the destruction and the millions of deaths that were happening around the world. I only knew that my dad had chosen to look away rather than help and I started to hate him a little bit in that moment. I had no idea that in a few months, I would be the one looking away from the suffering of the world.

My parents spoke softly in the front seat but I ignored them and searched my pack for my iPod. I wanted to drown out everything that was happening but when all I got was a blank screen I realized that I wouldn't be so lucky. Whatever had made all the cars crash had killed my small music player as well. I dropped it to the floor and closed my eyes. There would be no escape.

The drive into the mountains seemed to take hours and I dozed off now and then. I was half asleep when my mom let out a despairing moan. My eyes flew open and I searched frantically for what had hurt her. She was pressed against the passenger window, her palm flat against the glass, like she could reach out and change what we were driving past. I followed her line of sight and my breath caught at the view. A tangle of cars was crashed together half in the ditch, but what was so heart wrenching was the two figures standing beside them. I unbuckled my belt and slid over to the back passenger window so I could see them better. The boy was my age or a year older. It was hard to tell with the dirt and blood smeared on his face. In his arms was a wailing toddler that couldn't be more than two years old. I searched the crashed cars for any adults but it was only the two boys standing. I could hear my Dad chanting "I'm sorry" over and over under his breath as the truck slid by the pitiful figures. When our eyes met through the window I unconsciously mimicked my mom by raising my hand and pressing my palm against the cool glass. The older boy's forest green eyes stayed locked on mine as we moved past and at the last second he raised his own hand in a half wave and then they were gone out of view.

Mom was sobbing quietly in the front seat and Dad tried to comfort her but she batted his hand away. I had never seen anything but love and kindness between them before and it was just one more thing to absorb in this new world. His voice was empty when he started to speak.

"We have to take care of us now. We only have so many supplies and there won't be any way to get more. If we start taking in others we won't last very long. This is going to get a lot worse and there will be a lot more people who will need help. We just can't…" He trailed off when his voice cracked and Mom turned her head to look at me before facing him.

"They were just babies. What if it had been Skylar and our son out there? What if they were the ones all alone and hurt and people just drove past them? What if it had been our babies?"

Sitting behind my Mom I could now see Dad's face in profile. I saw his mouth tremble before firming into a hard line and his tone was just as hard.

"Every meal we feed to someone else is a meal our kids won't have in the future. I'm doing this ***for*** our babies. I'll do whatever it takes to keep you all safe. I'm sorry Van, that's just the way it's got to be now."

I slid back over the seat behind him and buckled back up. I didn't want to look at his face. I closed my eyes again and thought about how grateful I was to have a dad that would do anything to protect me and then I thought about those two boys who might die because I had a dad who would do anything to protect me. He wasn't the man I thought he was. I was so scared and confused that I didn't know how to feel anymore so I didn't. I just let it all go and slipped into a nightmare sleep.

The harsh smell of smoke woke me and I rasped out a cough to clear my throat. When the truck cleared the smoke, I saw that we were driving through the town of Canmore. We had spent a few weekends in the pretty little town that boasted it was the gateway to the Rockies. My family would stay at one of the many chalet style hotels and visit the tourist sites nearby and drive the short distance to Banff for ice cream and museums. There was a huge artist's community that my mother loved and we would always have to tear her away from the galleries in the small mountain resort town.

The highway ran through the town but each side was lined with a fence to keep tourists from trying to cross it on foot so it was free of people. There were crashed vehicles that Dad had to navigate around, causing him to slow the truck down enough that we had plenty of time to see the state of the hotels and businesses that lined the road. Almost everything looked the same as the last time we had been here except for one huge hotel that was being consumed by flames. There were people in parking lots and walking down the service road but I didn't see any vehicles moving at all. The sound of our truck's motor was like a siren and almost every person in sight turned their heads in our direction. When a few men started running towards the fenced highway and waving their arms at us, I felt the truck speed up. I had to brace myself when Dad swerved around a wreck at the higher speed but in seconds we had passed through town and it dwindled in the distance behind us.

We had only driven a few miles from the town when Dad took an off ramp and exited the hi-way onto a secondary road. After that, I had no idea where we were as he changed roads again and again as we drove deeper into the mountains. Eventually, the pavement ended and the sound of gravel under the tires started to lull me back to a light doze. When the gravel ended the bumpiness of multiple ruts in a dirt track started to bounce me around the back of the cab. My bladder gave a painful twinge at the bouncing and it made me think of Mom. For the last few months of her pregnancy, she seemed to have to pee every ten minutes and I wondered how she was holding it on such a rough road.

I leaned forward between the seats to try and see her face but she had angled her body away from my Dad towards her window. When I put my hand on her arm to get her attention, I felt a flinch course through it.

"Mom, are you ok?"

I snuck a look at Dad when she didn't answer me but his face was unreadable.

"Dad, I have to pee and I'm sure Mom's got to go too. Can we stop somewhere?"

He didn't answer for a few seconds and I was about to ask again when he let out a tired sigh.

"Can you hold it for a few more minutes? We're almost there."

I sat back against my seat and nodded at him in the rearview mirror, as scared and confused as I was about what I had seen in the last few hours I was also really curious about where we were going. I was just a baby when my Dad was a soldier. Mom said the last time he deployed I was two years old and I screamed for him all through the airport when he left. That was his last tour and when he came back he started up a business as a construction contractor. Even though I envisioned a cold damp cave full of bats, I knew Dad would never bring us to such a place to live long term. He had made the trip to his cave at least once a month and sometimes more often for as long as I could remember so I knew that he would have fixed it up into something habitable.

A huge rut in the path made the truck lurch and Mom let out a low moan. I was really worried about her and let a heavy breath of relief out when Dad slowed the truck to a stop. My peanut sized bladder was screaming for relief but I knew Mom would be even more desperate, so I grabbed a box of tissue from the seat and scrambled out the door before opening hers and helping her down. We moved away from the truck behind some bushes and I tried to brace Mom so she could go pee, but she waved me off and braced herself against a tree. I spun away and found my own tree to balance against and did my business. I might be a mall girl at heart but I had camped with Dad a few times so I knew to bring the used tissue back to be burned later.

Mom was back at the truck before me and she passed me a travel-sized bottle of hand sanitizer after taking my tissues and putting them in the plastic bag Dad used as a garbage in the front seat. As I rubbed my hands together, I looked around and saw nothing but wilderness. Closing my eyes, I took a deep breath of the late summer forest smell and tried to let the stress of the last few hours go. Opening my eyes, I looked at my Mom's strained face and sighed. As crazy as this living in a cave business sounded to me, at least we would all be together.

Mom leaned wearily against the side of the truck and cradled her big belly with one arm. Her face was pinched and tired looking as she watched my Dad pull our bags out from the truck bed. A small mountain of bags grew and I hoped we didn't have far to carry them all. Once he had removed everything he pulled my backpack and a few of the shopping bags from the pile and tossed them in my direction.

"Can you handle these, Sky? I don't want your Mom carrying anything."

With a nod, I gathered the bags and stood watching as he shouldered both his bug out bags and struggled to right Mom's suitcase. The tiny wheels on the case would not be much help on the rough, overgrown forest floor. As he tried to right the case, Mom pushed off the side of the truck and reached for the shopping bags draped over my arms. I gave her a concerned look but she just shook her head and took the bags.

"Go help your father with the case Sky, I can manage these."

I could see the stress that lined Dad's face and the way his jaw was clenched tightly. He only nodded at me and shot a quick glance at Mom when I helped him get the case upright and we finally started moving. It was a ten-minute struggle to get to where we were going and we were all panting and sweating in the late summer heat when he finally came to a stop and dropped the bags.

I looked around and saw only more forest and a rock wall. Seeing no cave opening I assumed we were taking a break, so I dropped my own pack and leaned against the rock wall to rest. I watched my dad take the shopping bags from Mom and gently rub her belly while speaking to her in a low voice. He kissed her gently on the forehead and then turned and stepped towards me.

"Can you move Sky? I need to get the door open."

I looked at him in confusion but stood up and turned to study the rock wall behind me. There was no door. Rock, moss, dirt and a few weeds poking out from cracks, but no door. I gave him a weird look and waved towards the wall and stepped back.

He gave me a teasing smirk before reaching out and pressing twice on a rounded rock protrusion. My eyes flared wide as a small square of rock moved to the side exposing a keypad. He chuckled at my expression and ruffled my hair.

"It's a false front. Now we just need to pray it was protected enough to survive the EMP."

He pushed a series of numbers that I realized was my birthdate and then hovered over the enter button. He looked at me and glanced back at Mom before taking a breath and pressing it. I think we were all holding our breath for the few seconds it took for something to happen. I flinched and stepped back at a clunking sound and watched wide-eyed as a piece of the rock wall seemed to break and slid to the side. I wasn't sure what I expected but the empty room that was exposed wasn't it. The tree-filtered sunlight shone into the room so that I could see that it was about ten by ten and mostly bare except for a drain in the concrete floor and a few panels that might be closets.

Dad put his hand on my back and nudged me forwards.

"Go on in, it's just an airlock. The real door won't open until this one gets closed again."

I glanced back at Mom but she just gave me a tight smile and waved me forward, so I grabbed my bags and stepped into the room. Mom and Dad followed me in with their own bags. Looking around the small, dim room I gnawed at my lower lip with my teeth in nervousness. I didn't know what was going to happen next when Dad reached out and put his hand against a panel I hadn't seen before and pushed a button. A scream was clawing its way up my throat when the door we had come through slid closed, taking the dim light with it.

Mom's hand on my shoulder changed the scream to a squeak of fear as we were plunged into blackness. As soon as the door made the clunk sound I had heard before, red light filled the room. My Dad said something that I couldn't hear over the pounding in my ears but the computerized voice that answered him rang loud and clear.

"Welcome back Daniel Ross. Voice imprint authenticated. Scanning for radiation and other environmental impurities. Scan complete. Sterilization commencing."

My mouth was dropped open in shock and I was scanning the ceiling for where the voice was coming from when the soft red light changed to an intensely bright UV beam that blinded me. Starbursts flashed before my eyes as I rubbed away the tears that had immediately sprung up. I felt Mom rubbing my back in sympathy as she blasted Dad.

"A little warning next time Daniel!"

Just as my eyes started to clear, the light changed again to normal fluorescent lighting and the clunk sound came again, this time from behind me. Dad stepped towards me and lifted my chin.

"Sorry about that sunshine. I should have warned you about the lights. Let's go in and I'll show you our new home."

I had a million questions on the tip of my tongue but before I could ask the first one, I felt a whoosh behind me and spun around. Just like the outside door, a section of the wall had slid to the side revealing the next room. I took a tentative step towards it and found myself snorting out a spontaneous laugh. My preconceived notions of a damp dark cave were swept aside as I took in the exact replica of our living room from the home we had left just hours ago. The furniture was the same as well as the family photos on the walls. The only difference was the paintings that were hung were copies of pieces my Mom had created instead of the originals.

I walked deeper into the room and saw open doorways to my left. Glimpses of beds and dressers showed me where we would be sleeping. To my right, past the living area, was a modern looking kitchen with an island counter and shiny new appliances. I spun around with a smile of delight that soon faded as I watched my Dad help Mom lower herself to the couch. Her face was strained and I immediately went and sat beside her.

She clutched my hand in hers and gave me a tired smile. "I'm ok, just tired. I just need to rest for a bit and I'll be fine." She tried to look reassuring.

Dad studied her face intently before finally nodding. He blew out a breath and rocked back onto his heels before speaking to us both.

"OK…there are a lot of things I need to explain to you both but most of it will have to wait for now. Time is of the essence in order for me to get everything we need to survive comfortably before we have to seal up and stay put. This will be our main living area. There is more to the cavern system further back into the mountain but I will show you that later. The voice you heard in the airlock is an artificial intelligence response interface application. Shortened down, its name is

AIRIA and it will answer questions and guide you in any way it can. So first it needs to voice print you both." He held up his hand and addressed the computer. "AIRIA, voiceprint my daughter, Skylar Ross. Access level yellow."

"Welcome, Skylar Ross. Please state your name and birthdate for voice print recognition. Access level yellow."

My voice trembled slightly as I said my name and birthdate. Mom did the same except she was granted access level green, whatever that meant. We sat in silence for a minute waiting for Dad to explain more of what was happening but he just looked at us uncertainly and rubbed at his face. I had never seen Dad look uncertain before and I was starting to get nervous again at what he might say next. He finally knelt in front of us and his face softened.

"Everything you need for now is in this room. You are both completely safe but there are things we will need in the future to make our lives more comfortable. I have to go, now, to get those things before the fallout makes its way to us and it will be unsafe to travel anywhere."

"Dad, NO…." I got out before Mom cut me off.

"Daniel, you can't leave us here! What if something happened to you?"

He gripped our hands tightly and gave them a little shake to silence us before explaining.

"Listen, I have planned this moment for a long time and I know what I'm doing! There's a farm nearby and I've had a deal with them for a few years. I pay to keep some livestock on their spread. We need those animals! I'll only be a few hours to get them loaded on the trailer I have there and get back here. This living area is only a small part of the cave system I've set up. There's another entrance closer to the back where I'll unload them and store the truck…I promise I won't be long!"

Mom was shaking her head as he spoke and burst out as soon as he finished. "How do you know it's safe? There might be radiation out there right now!!!"

Dad's mouth firmed up into a tight line before addressing the ceiling. "AIRIA, estimate time of fallout reaching this area."

"Yes Daniel Ross, calculating now. Based on the last satellite images available of the closest ground strike and prevailing winds, radiation contamination will reach your current location in seven hours and forty-two minutes."

Mom shook her head in confusion. "Daniel, how on earth could that computer know that? How do you have access to satellites?"

"Bill…well, let's just say this place is a backup to an official bunker the military built in the east. He wanted something out west just in case. We worked together on this place and he supplied the technical aspects of it as well as, well, almost everything."

Mom's mouth was dropped in shock. "Are you telling me this is a government bunker? Are soldiers going to show up here?"

Dad shot a quick guilty glance at me before focusing back on her. "Umm, no. This place is what you would call…off the books. Only Bill and a few trusted colleagues of his and us know about it. He used foreign contractors to build most of it and haul in the containers. When Bill called to warn me he was in the capital so I doubt anyone is coming."

He leaned forward and kissed her softly before turning to me and kissing my forehead. "Take care of your Mom. Help her get settled in and unpack some of the stuff we brought. I'll be back before you notice. If either of you need help just ask AIRIA and she'll help in any way she can. I love you both so much. I promise we will be safe here and I'll be back as soon as I can." He rose to his feet and looked down on our scared and confused faces. "I just need to grab a few things and then I have to go."

I sat holding Mom's hand and watched him stride over to another panel on the wall and punch in a code. Once again a section in the wall indented and slid to the side. He stepped through and disappeared. When the panel stayed open I couldn't help my curiosity so I squeezed Mom's hand and bounced up and crossed the room to the opening. I stood in dazed shock as I watched Dad's back as he walked further into a massive cavern. Lights sprang on ahead of him as he moved, exposing the huge room. To my left was a row of metal storage containers of the type that I had seen before on trains. I could only imagine what they were filled with. To my right was wire and mesh racks with garden containers on them. Below them were raised garden beds filled with soil but empty of plants. I wasn't ready to venture any further into the room but it looked like animal pens were set up further away. I marveled at everything that I could see and wondered when Dad had the time to set all of this up and how Mom and I didn't know about it. A soft moan came from Mom making me quickly move back to her side.

"What is it? Is the baby ok?" I asked in fear.

She shook her head and took deep breaths before answering me.

"I'm ok honey. It's just the stress of all this. Can you help me stretch out? My back is killing me. Then maybe you could get me some water."

"Of course Mom, just rest!"

I helped her get her feet up onto the couch and pulled a soft afghan from the back of it to cover her with. Patting her on the shoulder, I headed to the new kitchen and started to open cupboards. I had to shake my head in wonder as I saw replicas of our dinnerware from home. Dad had gone to so much trouble to make everything just like home that I felt a swell of love fill me. After checking out the pantry and finding boxes of shrink-wrapped food I grabbed a glass and opened the fridge. Inside it had a shelf full of bottled water and another one had stacks of different sodas. I grabbed a cold bottle of water for Mom and a can of Ginger Ale for myself before going back and settling on the floor beside her.

Her soft breathing told me she had dozed off so I left her water on the coffee table and sipped at my soda while thinking about what all had happened so far today. I started to get overwhelmed at the thought of never going back home or school or anywhere that I was used to, and I felt tears start to well in my eyes. I just couldn't believe that my world was ending. Looking around at the copy of my home just made me more sad, so I closed my eyes, leaned back against the couch and waited for whatever came next.

A soft hand on my shoulder startled me from my deep thoughts. Dad was looking down at me with eyes that were filled with worry and love.

"Sky, I have to go. I need you to help Mom get settled in. I promise I'll be back in a few hours." He paused and brushed a stray curl from my forehead. "Everything is going to be ok, I promise."

I opened my mouth to reply but nothing came out so I just nodded and pushed to my feet and leaned into him. His strong Dad arms wrapped around me and made me feel safe but all too soon he gently pushed me away and leaned over Mom. I stepped away to give them privacy as they whispered words to each other and then he was gone.

I stood staring at the door he had left from for at least five minutes before turning back to Mom. She had closed her eyes once again so I grabbed a few bags and carried them into what would be our bedrooms. One of the rooms had a huge king sized bed so I guessed it was my parents. I nudged the other door open and peeked in. A double bed was against one wall with a desk beside it. Sitting on it was a small radio with an iPod resting in the docking station. A few long strides had me across the room reaching out to thumb on the player. A smile cracked my face as it lit up and displayed all the playlists I had on the dead model I had left in the truck. Shaking my head in wonder at my Dad's attention to detail, I spun around the room and took in the Taylor Swift and One Direction posters on the wall.

The smile slowly faded from my face as I realized that those music artists may no longer exist. I dropped the bags on the bed and made my way back to the living room. I couldn't help but start to think about how many people might now be dead. The computer voice had said bombs but how many and where?

Looking over at Mom and then scanning the ceiling I tentatively asked, "Umm, Computer? Are you there?"

"Skylar Ross you may refer to me as Computer or AIRIA. How may I assist you?"

"Umm, ok, AIRIA you can call me Skylar. I, uh, I just wondered if you knew how many bombs have gone off and um, if very many were close to here."

"Skylar Ross, last satellite data was from thirty-six minutes ago. At that time there had been two thousand sixty-three ground strikes with three hundred and thirty-nine occurring in North America. The closest strike was Edmonton, Alberta. The second closest strike was Vancouver, British Columbia. Up to date satellite imagery may be obtained in the future if undamaged satellites come into communication range."

A gasp from Mom had me meeting her horror filled eyes. She had stood up from the couch and her hand reached out to take mine. Her eyes had a sheen of tears in them and her voice was a harsh whisper.

"My God, how many people must have died?"

"Vanessa Ross, estimates of initial casualties from ground strikes would be three billion. Calculating secondary casualties from radiation fallout and infrastructure breakdown now..."

"NO! I, we don't want to know that!" Mom was shaking her head frantically. "Oh, oh Sky, half the world just died!"

The tears were pouring down both our faces as we tried to grasp the huge number of people who had just died when without warning Mom's grip on my hand turned bone crushing and she doubled over with an agonizing moan.

"Mom! Mom, are you ok? What is it?" I tried to help her back up but she just kept moaning until she fell to her knees.

Her breath was a harsh pant as she raised her head and pain filled eyes met mine.

"Sky, the baby...my water just broke!"

Chapter Two-Rex

"Rex, honey, can you check on Matty please?"

I hear Mom ask me even over the sounds blasting my ears from the game coming through my earphones. My eyes lift from the DS gaming system in my hands for a split second to meet hers in the rear-view mirror before cutting across the back seat to my baby brother who is sleeping in his rear facing car seat and then back to the screen and my game. I grunt out that he's fine before tuning anything else she might say out like only an eleven-year-old boy can.

I have no reason to speak to her about anything. She didn't speak to me when she made the decision to get divorced from my dad or to move us away from him and all my friends to a totally different province. Nope, nothing to say.

It wouldn't be until years later that I put scattered pieces of overheard conversations together to realize that he had left us for someone else and he didn't want to pay child support, making Mom lose the house and forcing her to move to Alberta so we could live with my grandparents. All I knew that day was that she was ruining my life and I hated her for it. The only thing I could do to punish her was to ignore her. It was the worst day of my life - then it really got bad.

I squawk in outrage when my game player goes black and with the silence of the game comes the silence of the van. Something's wrong. We were still moving but it feels like the van has turned off. I lean forward in my seat and see Mom fighting with the steering wheel and jamming her foot against the brake pedal but we're still flying down the road. My eyes grow huge as I see the cars ahead of us crash together and I know we're going to hit them.

"Mom, Mom, Mom STOP!!!" Screaming was the last thing I remember before the impact and with the pain came blackness.

Silence, absolute silence. White flashes across my vision and something's burning in my eyes as I try to open them. I struggle to lift my arm up to rub at my eyes but it feels so heavy. Finally, I get it up and rub at what I think is water blurring my vision until I can open them. Blood, not water, blood on my hand and in my eyes. Then the pain…ahhhh…it hurts so much! Blood, pain and silence, oh my God why can't I hear anything. Look, look around, find Mom, find Matty. Oh man, are we sideways? Everything is tilted. Can't see Mom. Matty, Matty is there! He's almost underneath me but still in his seat and his mouth is open and I know he's screaming cuz his eyes are pinched shut and his face is red with tears and…SOUND!

So much noise, too much for my aching head! Matty is screaming and metal is creaking and people somewhere outside of the car are yelling and moaning and crying but where? Where is the most important noise? Where is Mom's voice? Where is the most important voice in my whole world asking if I'm ok or telling me I will be? WHERE IS MOM???"

I pat at the seat belt and look down around me and see that I will fall right on top of the baby if I unbuckle so I brace my feet against the seat in front of me and hold on to the armrest on the door before hitting the release button. Ahhhh, that hurts, everything hurts! But I hold on and pull myself up to the front seat and slide through them so I'm anchored. My breath whooshes out in relief, there she is. She's against the door and her long brown hair is covering her face but it looks like she might have been knocked out like I was. I'm calling for her to wake up and pulling at the one arm I can reach but she's not waking up so I slide further into the front of the van until I can just reach her hair with my fingertips and I pull it away from her face.

Finally, she's awake! She's staring at the dash in front of her even though I keep calling her name and I think for a second it's payback for me ignoring her but I know it's not. I know what it is but I can't, so I push it away and I beg because I need her to look at me. I need her to hear me when I say I'm sorry and I need her to say it will be ok but I know that will never happen now so I let it all go away again.

Whimpering, someone is whimpering and it's so sad, then a hiccup that turns into a cry and then Matty is wailing full force again. My eyes clear and I take one more look at her beautiful face before brushing her hair back over it. I pull myself back into the back seat and ignore the pain and ache in my chest to get to my baby brother. He has no idea what has happened and I have to help him. I'm all he has left now.

I anchor myself against the seats and lean over his car seat. Matty's face is tomato red and wet with tears and snot but he doesn't seem injured on the surface. I rub his head and pat his leg until he opens his eyes and focuses on me. He whimper's my name but in baby speak it comes out "X", poor kid hasn't even learned to say his r's yet. I speak rambling reassurances to him until his tears become stuttering hiccups. I have to get him out of the van but the way it's tilted will make it a huge challenge.

I look past him and see long grass and weeds pressed against the van's window so I know we won't be going out that way. I also see his Sippy cup lying on its side near the roof so I grab it and offer it to him. His chubby fists grab onto it like a lifeline and he stuffs the spout into his mouth. I take another look around the back seat and spot Matty's diaper bag so I snag the strap and pull it up and stuff it in between the seats so I won't forget it. Looking up at the side door above my head, I reach up and try to pull it open but it's stuck, locked. Stupid child locks mean I'm going to have to go back into the front seats and unlock them.

I close my eyes and try to roll the pain out of my shoulders and neck but the ache isn't going anywhere. For the next five minutes, I wiggle and squirm in between seats trying to get the doors unlocked and open. I force myself not to look at Mom and take deep breaths to push the pain and devastation back so I can get Matty out. No matter how many times I hit the unlock button nothing works. I'm going to have to break a window.

I sit for a minute and try to figure out how to do that. I turn and look at all the van's windows that are all intact. I'm not strong enough to break any of them. The only shot I have is the windshield. It's still in place but there are cracks all through it and some of it is webbed in hundreds of little cracks. I want to kick it but the way the van is tilted I can't stand up or lean on anything to get my leg up to kick without falling down on Mom. Tears of frustration build in my eyes. Where is everyone? Why hasn't anyone come to help us yet? I know there were other cars that crashed ahead of us and we hit them but didn't anyone survive?

"X, X, X…Maaaaaa?" Comes from Matty in the car seat and I can tell he's amping up again for another wail so I slide back between the seats and lean over him.

"I'm here Matty. I'm here! I'm going to get us out of the van, ok? Just hold on and I'll get us out of here."

His fat little bottom lip trembles and he pats at my face nearly poking me in the eye so I give him my best smile and pretend I'm going to bite his fingers which makes him laugh. "Be right back, Buddy!" I fake a cheerful tone and move away to the back of the van.

The seats back here have been put down to make room for all the stuff we didn't send ahead with the moving truck. It's mostly suitcases of our clothes and garbage bags filled with blankets. Matty's monster stroller and toys are in a scattered tangle. None of this stuff is going to help me bash the window out so I start shifting things, looking for anything I can use for a battering ram. I grab a small stroller wheel and pull it to the side before realizing it's from the small umbrella stroller Mom uses when she doesn't need the big one. It's light and folded up, all the wheels come together. I might be able to use it so I pull it closer before looking for something else to use before giving up.

The back window of the van is now clear and I can see out. We're in the ditch and there's nothing behind us. If I can get out of the windshield I can use the keys to unlock the van's rear hatch and get Matty out that way. I slide the small stroller through the seats and up into the front before tweaking Matty's nose to make him smile and then follow it.

Before I start pounding at the glass, I get the keys out of the ignition and stash them in my jeans pocket. There's a strap that goes around the stroller when it's folded up to keep it from opening and I make sure it's snapped closed before aiming the four wheels at the window. My first strike is tentative and does nothing. I take a stronger hold and give it a harder push. Some creaking and cracking noises but it doesn't shatter and fall like I want it to so I start really hammering on it. I put all my anger, frustration and fear into it and on the fourth hit the wheels go straight through the window. The stroller gets hung up half way in and half way out of the windshield so I wiggle it back and forth until I can pull it back through and keep on pounding. I finally get a hole just bigger than my head when I pause to rest. Sweat is dripping down my face mixed with the blood from the cut on my forehead making my eyes burn again. There's a box of tissues in the center console so I grab a wad and wipe away the mess and hold it to my cut wincing at the sting of it while deciding what to do next.

Matty had started wailing at some point but I just ignored him. All I want to do is close my eyes and go to sleep. Everything hurts and I'm hot and tired. I shake my head; have to get the baby out. With all the windows closed, the van is quickly heating up in the late summer sun. I pull the stroller back towards me from the hole I had made and see how the glass sort of bends with it. Dropping my makeshift ram between the seats, I used the wad of tissue to grab the edge of the hole. I'm scared I'm going to get my hand sliced open from the glass but it doesn't seem to be sharp. I give a tug on the edge and the glass bends towards me. Grabbing more tissue for my other hand I start to bend the glass back and forth until part of it comes away from the frame. After that, I stand on the console and used my shoulder to push at the window until it folds over out of the way.

My head and shoulders clear the frame of the windshield into the fresh air. I take a deep breath and choke on the awful smell of chemicals and gasoline. The smell of gas is like a firecracker under my butt. I have a vivid image of the van exploding with the baby still inside. Our van is on its side with another car embedded in the hood. I stand on my tiptoes and try to clear the rest of the glass. I expect to feel the glass cut into my belly as I press against it for leverage but it just bends and breaks away sending me down onto the trunk of the other car. I can just make out a person's head in the front seat of the car we hit but the blood covering the windshield makes me gasp and turn away. I barely catch my breath before I'm scooting off the trunk and down to the ground. I stumble around the van on the uneven side of the ditch and make it to the hatch. The keys work just fine unlocking it and the door comes up with no problem. Finally, something is going my way.

My neck and shoulders are screaming in pain as I haul out the suitcases and bags. The monster stroller almost makes me sit down and cry. It keeps getting caught on stuff before I finally get it out of the hatch and out of the way. I'm so tired and thirsty, I just want to sit in the shade and go to sleep but Matty's still crying. Using the side of the van and its windows as my floor, I walk in only slightly hunched over. The first thing I do is grab his diaper bag and toss it out the back. I finally get the car seat buckles undone and keeping the poor little guy braced so he doesn't fall against the window, I untangle his flying arms and pull him over the seat and against my chest. When his small little arms wrap around my neck and his hot, wet face buries into my shoulder, I almost lose it. My own chest is hitching with silent sobs as we climb out of the van.

The urgency to get away from the van fades as I stand in the ditch with Matty clutched against me. The slight breeze cools the sweat on my face and we just stand there for what feels like an hour. My feet finally move of their own accord and I climb the bank of the ditch. I need to find help but what I see up on the road freezes me in place. Wreckage, in both directions, nothing but crashed cars and trucks. Nothing is moving and there's no sound but the wind in the trees and the tick of hot metal cooling in the wreck.

I don't know what to do. I got us out of the van and now someone's supposed to help us. I'm only eleven, I need an adult to help us. My mind just sort of shuts down after that and we stand there beside the road waiting for what comes next. I don't know how long I stood there but my arms started to ache from holding the baby's weight and my throat was desert dry. I glance back to the ditch and think about moving to find some water and shade to wait in when something changes. The silence is faintly broken by the sound of an engine and it's getting closer. Finally, someone's coming to help.

I stand there and wait as the sound draws closer. Matty even stops crying and turns his head in the direction it's coming from. A pickup truck drives slowly towards us as it makes its way around the accidents that litter the highway. My baby brother and I watch it get closer and closer and my knees almost give out at the relief of someone to help us. It comes close enough that I can see a pretty woman with a horror-filled face staring out at us and as it starts to pass us, she raises her hand and presses her palm to the window. In the back passenger window, the face of a girl appears and she has the saddest eyes. She raises her hand to the window just like the woman in the front had. My hand comes up to mimic theirs in a small wave as I wait for them to pull over and stop. My hand hangs in the air and I watch in disbelief as they just keep going. When I can no longer see the truck my hand drops and my knees give out. I don't feel the pain of the hard pavement as my knees crashed against it. The pain of being abandoned eclipsed everything else and I finally weep.

I don't know how long I kneeled there on the hot road but the back of my head is on fire from the sun and Matty's patting my leg and babbling at me when I come back to the world.

"Kay X, kay X? Dink X, dink?" was what I finally make out of his baby talk. I try to tell him that yes, I was ok but my mouth and throat were bone dry and all that came out was a groan. After rubbing the tears from my face, I painfully rise to my feet and haul him up onto my hip. I have to get us out of the sun and get some water into us.

I stumble down into the ditch again and sit Matty in the grass. Shoving aside some of the bags I had removed from the back of the van, I finally unearth his monster stroller and get it unfolded and upright. I pull the sun shade canopy open and plop him in it before finding his diaper bag and getting drinks for us both. Once I can swallow without choking on the dryness in my throat, I go back into the oven the van is becoming and find my backpack. With my Saskatchewan Roughriders cap on my head, I scan around the mound of bags and try to decide what to do next.

Matty has fallen asleep in his stroller so I climb back out of the ditch and look both ways. Still nothing, I don't understand why there aren't more cars driving through. With this many accidents, there should be ambulances and police cars everywhere by now. Shaking my head in confusion, I go back down and start to sort out our belongings. When emergency crews finally do show up, they will take Matty and me somewhere to meet my grandparents and we'll need a few things.

The first suitcase I open brings another sob to my throat. The smell that wafts out brings comfort and home. It's Mom's smell. Every hug, every cuddle, every soft touch comes with that smell and it now brings a sharp arrow of pain to my chest. Looking down at her clothes through a haze of tears brings her loss full center again. I have no choice but to gently close and re-zipper the bag and move on. There would be plenty of time later to think about her.

I get everything I think we will need into one suitcase and drag it up the ditch closer to the trees where a slice of shade has grown. Once I pull Matty and his stroller up and into it, I sit down and close my eyes to wait.

I have no idea how long I slept or what woke me up but the slice of shade we were in has grown and moved almost to the road. I peek over the side of the stroller and give a sigh of relief when I see Matty still sleeping. My stomach grumbles for food so I guess it's close to supper time. Pulling the diaper bag open I start to search for something to eat when I hear singing. My head whips up and I search for where it's coming from. The voice sounds like it's coming from a girl and it's getting closer. I scramble to my feet and rush down and then back up the ditch until I'm standing on the road.

I stand with the sun at my back and watch as two figures skirt around the closest wreckage. It's a tall slim woman with soft red hair that's piled on top of her head. Holding her hand is a girl a few years younger than me and she has the exact same colour of hair but it's loose down her back and it floats behind her in the light breeze. The girl sees me first and her song cuts off mid-word, alerting her mom who looks up and shades her eyes against the sun.

I swallow past a lump in my throat and raise my hand in greeting. The voice that comes out of me doesn't sound like mine. It sounds years younger.

"H-hello?"

They walk a few steps closer before stopping and the woman tilts her head and studies me for a moment before speaking.

"Hello sweetheart, are you all alone?"

I start to say yes but then remember.

"No, my brother is over there in his stroller. He's just a baby."

They both look over at where I had pointed and then she asks, "Your parents?"

I try to answer but my throat closes up in sorrow so I just shake my head and look down at my feet. I hear them walking closer and then soft arms are around me and a hand is rubbing my back and it all pours out. The physical pain from the accident, the anger at moving away from my life, the guilt at treating her so bad and those empty staring eyes. I cry every tear in my body against that soft chest and she takes it all.

When I'm finally empty and still, she pushes me back a bit and wipes my face with her hands and says the words I have been waiting to hear.

"It's going to be ok."

That almost sets me off again but I bite the inside of my cheek to stop myself from wailing again and step back from her arms to bend down and grab my cap that has been knocked to the ground. I anchor it on my sweaty head and glance around to see if the girl had seen me crying like a baby. I spot her across the ditch leaning over Matty's stroller. She stands up, turns back our way and waves us over.

"Come on sweetie; let's go check on your brother. By the way, my name is Belle and that's my daughter Sasha. What's your name?"

We start to walk down the ditch when I tell her mine and Matty's name.

Sasha is super skinny with long arms and legs but she has no problem scooping Matty up from the stroller and holding him out to her Mom.

"He looks like he's alright Mom but he really stinks!"

Matty waves at us with chubby arms and a big grin and announces, "X, X, poopy, poopy!"

I groan in dismay at the thought of changing his diaper. The kid's a poop bomb master but Belle laughs at him in delight and swoops him into the air out of her daughter's arms.

"Hi, hi Matty! I'm Belle and I'm going to change your diaper for you. Is that ok?"

The kid's response was to pat her face with a satisfied grin.

Belle takes over with no hesitation by kneeling by the diaper bag and pulling out all the supplies she needs. She has the kid poop free and clean in minutes. The ache in my chest lessens a bit at having an adult here to help us.

Now that he's clean and free of his stroller, Matty starts to crawl around and pull out grass. I let him explore for a bit while I help Belle and Sasha pull out the little bit of food I have taken out of the van. They have a few things to eat from the backpacks they had been carrying, but it's not much. As we eat our meager supper Belle explains what they had been through.

"I don't know what caused it but the car just shut off. Everything went dead at the same time. I think the only reason we made out ok was because I had slowed right down before it happened when I saw some elk beside the road. Every other car near us had to have been doing a hundred kilometres an hour and it would have been nearly impossible to stop safely. The power brakes were useless so I just used the emergency handbrake as much as possible until we finally stopped. We tried to help the people in the accidents closest to us but almost all of them didn't make it. The ones who did just sat and waited for help to come. We waited by our car for a few hours but no one came so we decided to just start walking. My best guess based on the sun is that it's been at least three hours since everything stopped working. Something big must have happened for emergency services to not have responded by now. With all the cars and cell phones dead, I think we must have been hit with some kind of terrorist attack. The only thing I can think to do is walk back to Canmore and try to find help. I think you and your brother should come with us."

I look over at Matty who's stuffing animal crackers into his messy face and sigh. We have to go with them. There's no way I want to sit here and wait for someone to show up. I just can't bear to stay here with Mom's body so close.

I nod my agreement and we start to gather up what's left of the food. Belle is so good with the baby. She has him wiped clean and in his stroller with some of his toys in less than half the time it would have taken me. As Belle and Sasha make their way up onto the road, I keep glancing back at the van. It just feels so wrong to leave Mom there all alone. I just don't know what to do but I feel like I need to say something or do something. I finally hold up a finger in a just a minute gesture to Belle and climb back inside the van. There's a faint sour smell inside that gets worse as I climb between the seats to the front. I brace my feet and lean against the passenger side seat.

Her hair's still covering her face and I want to brush it aside to look at her one more time but I'm scared she wouldn't look the same. Instead, I pat her cold hand that rests in her lap.

"Mom, we have to go now. No one's come to help but a nice lady and her daughter are going to walk with Matty and me towards the last town we went through. I'll take care of the baby, Mom. I promise." I sucked back the tears that wanted to clog my throat and say the one thing that was weighing the heaviest on my heart. "I'm sorry, Mom. I'm so sorry I was such a jerk to you. Mom…Mommy, please come back. I'll be better, I'll listen and help out more. Just come back…please?" In my head, I knew that wasn't possible but my heart was begging for it to happen.

After a moment of silence, my shoulders slump in defeat. She isn't coming back and I needed to say goodbye. Bracing myself on the side of her seat, I lean over as far as I can until my lips could just brush her hair.

"I love you." choked out of me. As I opened my eyes and begin to pull away I catch sight of her purse leaning against her feet on the floor. Reaching down, I snag the strap and carry it out of the van with me.

Up on the road, Belle gives me a compassionate smile and rubs my back.

"Ready to go?"

As I nod my head wearily, the sound of an engine breaks the silence around us. We all turn to look with hope in our hearts. Hope leaves me just as quickly when I recognized the dark green pickup truck headed towards us. It's the same one that had passed me and Matty earlier, although this time it has a long black trailer behind it. I turn away from it to Belle.

"Don't bother, it won't stop."

She gives me a weird look and starts to wave. When the truck pulls to a stop beside us there's a scowl fixed on my face as I look for the woman and girl that had stared at us earlier. The truck's empty except for a man who gets out and walks around the front of it to get to us. I have no way of knowing that mine and Matty's faces have left a haunted wedge of guilt in the man's heart.

Belle's practically gushing at the guy before he even makes it around to the front of the truck.

"Oh thank God! Thank you so much for stopping! We were just about to start walking towards town."

The man looks Belle, Sasha and me over before stepping closer and glancing into the stroller before nodding.

"Right, get in quick and I'll put the stroller in the back. We have to hurry! I'll explain what's going on once we're on the way."

We all just stand there and stare at him until he barks out at us.

"If you want a ride to town then get in! We don't have much time until the radiation gets here!"

At the word "radiation", Belle's face goes completely white and she practically throws Sasha towards the truck before snatching Matty from the stroller. The man grabs the suitcase and tosses it in the bed of the truck and tries to figure out how to collapse the stroller for a few seconds before giving up and heaves it into the back on its side. I still stand on the road with my scowl, not trusting the guy, until he looks my way. His lips are pressed grimly together and with a quick nod at the truck he says, "I'm sorry but I came back for you, now get in!"

I opened my mouth to say something scathing to him when a very distinctive sound comes from behind me. I slowly turn around and look at the long trailer hooked to his truck before asking in disbelief, "Is that a cow?"

Chapter Three-Skylar

Mom was doubled over and groaning in pain as I stood like an idiot and watched the area rug absorb the puddle that had formed under her feet. I didn't realize I was shaking my head in denial until she tugged on my crushed and throbbing hand.

"Sky, I need you to help me! Go get some towels from the bathroom and see if you can find some plastic garbage bags in the kitchen."

I just stood staring at her with my mouth wide open until she groaned again and gasped out, "NOW SKY!"

I ran to the bathroom and gathered up all the towels I could carry. My brain was on a chanting loop that went, "NOT HAPPENING, NOT HAPPENING, NEED DAD, NEED DAD" over and over again. There was no way I could help Mom deliver the baby so I just prayed that Dad would be back in time to help her.

I dumped the towels on the couch and raced to the kitchen where I started throwing open cupboard doors in my search for garbage bags. I think I opened every door before I finally thought to look under the sink, where of course there was a box and sitting right beside it was a large first aid kit, so I grabbed it too. A groan of pain from Mom sent me back to the living room leaving the kitchen looking like a tornado had gone through it.

Mom was still bent over clutching her huge belly with one hand as she tried to spread out towels on the floor with the other. I dropped the bags and kit and tried to guide her to the couch but she pulled away and shook her head with heavy pants.

"No, on the floor. Put the plastic down first and then towels on top of it. I'll need the hard support of the floor when I start to push."

I started to nod and grabbed a few plastic bags to spread when what she said penetrated my panicked thoughts, making me freeze and gape at her.

"Wait, what, push? No, no Mom, no pushing, you can't push! You have to wait for Dad to come back!"

Her teeth were clenched against the pain so her smile and chuckle came out slightly demented which didn't help reassure me at all.

"Babies don't wait when they're ready, honey and this one is definitely ready to come out! The shock and stress must have triggered my labour. I'm guessing it started in the truck on the way here. I just thought it was stress."

Her shoulders came down and the strain on her face cleared as she took a deep breath and let it out with a whoosh as the contraction ended.

I was still standing with my mouth wide open and my face white with shock holding out the garbage bags when she focused on me. Her expression softened and she reached out and cupped my face.

"Oh, honey, it's ok! It's going to be fine! This isn't the first time I've done this remember? You were such an easy birth, six hours of labor, forty minutes of pushing and voilà, there's baby! Second babies usually come faster so this should be quick."

Her tone was reassuring and she was trying to make light of it so I wouldn't be afraid, but I could see the fear in her eyes reflected back at me. She took the bags from my hand and we worked on spreading them out over the rug before covering them with some towels. As she started to lower herself to the ground, another contraction gripped her and she went to one knee with the loudest moan yet. I had tears streaming down my face at how helpless I felt. I didn't know what to do or how to help her. I kept glancing at the door in the hopes that Dad would walk through it and take charge so I could go and cower in my room. It was breaking my heart seeing Mom in so much pain, even when she reassured me that it was natural and part of the process.

Every minute seemed to last forever when she was panting through a contraction and the breaks between seemed to get shorter and shorter as time went by. Mom had me open the first aid kit and lay out some of the supplies. She also sent me on a search for string that she said we would need to tie off the baby's umbilical cord before we cut it.

Again, my body was in motion before my brain caught up to the implications of her words and I was like, "Right, find string to cut the cord", and then it penetrated and my head whipped back and my stomach heaved.

"Cut the what? We have to cut something on the baby?!!!"

This was too much for me, I'm only ten, and I'm not supposed to know any of this let alone see it happening. My heart started racing and I was having a hard time catching my breath. My hands were fluttering in front of me and I didn't know what to do with them.

"Sky, SKYLAR! Look at me, look at me! You are having a panic attack and you'll hyperventilate if you don't calm down!" Mom yelled at me in her very effective Mom tone before grinding her teeth and groaning.

I don't know whether it was the no-nonsense tone or her pain that snapped me out of it but I dropped to my knees beside her and gripped her hand.

"I'm so sorry, Mom. I'm just so scared. All of this, it's just…we were going shopping for back to school clothes like five minutes ago and now we live in a cave and bombs are dropping and you're having the baby on the floor!" My voice got faster and more shrill as I spoke. She gripped my hand tighter and pulled me down beside her.

"I know, I know how overwhelmed you are, honey, I am too. This certainly isn't how I planned to have the baby and I'm trying to be grateful that we have a safe place that will protect us from what's happening in the world. The most important thing is that we're together and safe, so focus on that." She was going to say more but the next contraction hit so she just grunted.

I brushed her hair off of her sweaty face and got to my feet. Dad wasn't here so it was up to me to help her. I finished gathering what she had asked for and then went to the bedroom and emptied all the bags I had brought from the nursery at home out onto the bed. I sorted through it and found a tiny baby sleeper with a matching hat as well as washcloths, baby soap and the tiniest diaper I had ever seen. I carried it all out to the living room and left it on the coffee table before going into the kitchen and finding a big bowl to fill with hot water.

I almost dropped the bowl as I turned to take it to the table when Mom started making new scarier noises. Something had changed, her face was red with the strain and her groan was a drawn out sound of determination. I got the water safely to the table and kneeled beside her. Her groan ended with heavy panting that sounded like she had just done a mile-long sprint.

"Have to push! Baby's coming fast!" she gasped out.

The next twenty minutes were the scariest, most terrifying moments of my life. There was nothing I could do to help her except wipe off her sweaty face and mumble empty platitudes. We both had tears constantly falling down our faces as she rotated between screams, cries for my Dad and trying to reassure me that all this was normal for having a baby. I could only nod my head and try not to run away and hide. When she clawed her dress up over her knees and yelled at me to get a towel to catch the baby with, I had to force the vomit that filled my throat back. The baby came out with a gush of fluid and blood and I almost dropped him. He was covered in blood and sticky white goo. All I could do was stare down at him. Mom's laughter had my head snapping up to meet her eyes and I passed him to her outstretched arms. She cleared out his tiny mouth before turning him over and rubbing and patting his small back. When his angry cries rang out, I found myself bursting out with laughter too.

Mom gave me instructions on what to do next. Tying and cutting the cord was totally gross but not as bad as I thought it would be. After that, it was just cleaning the poor little guy off with the now lukewarm water and getting him diapered and dressed. Once I had him wrapped up in his sleeper and soft blanket, I turned to pass him back to Mom. Her face was serene but paper white. Her tired eyes met mine and a faint smile passed over her lips.

"Benjamin, his name is Benjamin. So proud of you Sky. Couldn't have done it without you. Love my babies so much." Her words faded away as her head nodded and her eyes fluttered closed. I couldn't keep the grin off my face as I looked down on the tiny sleeping baby in my arms until I felt something warm and wet touch my knees. Glancing down at the floor, it was an instant reaction to scramble back away from the pool of blood forming underneath Mom's legs.

I practically threw the baby on the couch behind her head so I could grab her shoulders and shake her awake.

"Mom, MOM, wake up! Wake up, something's wrong, you're bleeding! Mom, what do I do?" Her head flopped loosely as I shook her but her eyes stayed closed. I pushed towels against her to try and stop the bleeding but it didn't seem to help. I was sobbing her name when the computerized voice spoke from the ceiling.

"Vanessa Ross, no life signs detected."

I screamed out an agonizing wail that startled the baby awake and his cries joined in with mine. I don't know how long I spent crying and begging her to wake up but Benjamin's screams finally had me leaning away from her. I laid the rest of the towels over the blood and her legs before picking the baby up and settling the both of us beside her on the floor. We cuddled up against her and I held her hand between me and the baby and then we waited for Dad to come back.

Chapter Four-Rex

Without a word, he pointed at the door to the back seat so I turned and climbed into the truck beside Sasha. I heard a crunch and looked down to see that I had stepped on an iPod and it now had a shattered screen. Using my toe, I nudged it under the front seat out of sight. I didn't want the guy to have a reason to throw us out.

Sasha saw what had happened and her eyes were big and scared. She opened her mouth to say something but I gave a quick shake of my head to silence her. We both lurched back in our seats as the truck started moving so we scrambled to put our seat belts on. Belle was holding Matty in the front seat and was firing off questions to the man who was ignoring her. Once she ran out of steam he glanced back at us in the rear-view mirror and then began to speak.

"Nuclear war has broken out. Thousands of bombs have been dropped so far with more probably on the way. The closest ones that I know of are Edmonton and Vancouver. Nothing electronic works because of the electromagnetic pulse that goes out with those kinds of bombs. Radiation from Edmonton will head East with the prevailing winds but Vancouver's is already heading this way. I'm not sure how much will make it over the mountains but some will, and we all need to be under cover before that happens. I'm taking you guys to the supercenter in Canmore. I need to grab baby supplies and you need to find a basement or stock room to take cover in. If you and these kids want to live, you need to listen to me." He took his eyes off the road for a second to drill Belle with a look of seriousness. When she nodded vigorously he looked back to the road and continued.

"The looting shouldn't have started yet. Most people will still be waiting for help to come. The biggest threat will be people trying to steal this truck because it's still working. I've mapped that town out so I know how to get around and avoid the main streets. We will park behind the store on a service road and hope no one sees us. We go in through the back door if we can get it open and start grabbing stuff. You need enough food for all of you for at least two weeks. This is important! Once you find your room and get it stocked, you need to lock it or barricade it closed and stay in it for at least two weeks. No matter what you hear, you don't come out! Radiation isn't something you can see. You have to wait it out to be safe. Once you come out, make sure you are all covered up. There will be ash and it will be dangerous. Also stay away from metal, it holds radiation. Other than that it will be people that are the most dangerous. Food and clean water will be like gold and people will kill for it so be ready to protect yourselves. Don't trust anyone."

Belle was silently weeping as she listened to him but she had a look of determination on her face. I couldn't believe what the guy was saying as I looked around the sunny early evening scenery we were passing. Except for all the car crashes and the occasional person walking on the road that tried to wave us down, everything looked normal. The world this guy was describing just didn't make sense. His harsh voice broke me out of my trance.

"How old are you, son?"

I scowled at his reflection in the mirror before answering, "I'm eleven and I'm not your son!" I was still angry that he hadn't stopped to help us earlier and I let him be the target for all the emotion of the day.

"Sorry, but listen up! You have to help your Mom protect the smaller kids. She's going to need you to survive this!"

I glared at him and tried to scream that she wasn't my Mom. That my Mom was dead back in the car he had passed but nothing came out. His words about helping my Mom had sent a spike of pain and guilt through my chest.

Belle must have said something to him because he frowned and shook his head.

"Hey, I'm sorry kid, I didn't know." It got very quiet and we drove in silence for a while until Belle asked him what his name was.

He glanced at her and back at us before telling us, "Daniel Ross."

We left the highway just as signs for the town came into view and the guy made turn after turn on back roads until he finally came to a stop. We sat in the silence as the man squeezed the steering wheel and sat with his eyes closed and his head bowed for a minute. He took a deep breath and let go before turning to us in his seat. He looked at Belle and then looked back at me. He looked down at his watch and then addressed us.

"Ok, the store is just on the other side of those trees. Hopefully we won't have any problems and we can do this fast. I'm leaving that store in exactly forty-five minutes, no exceptions. Once I have what I need, I'll help you as much as I can but I have my own family I have to get back to. Don't wait for me; get to work as soon as I tell you to go." He nodded at Belle. "You, start with water. Grab a few carts and pile them high with as many jugs as you can." He turned to me. "You, go for food. Get cans of soup and ready to eat stuff, lots of cans of tuna and boxes of crackers and juice. Just think of easy to prepare stuff that will last." When I only nodded, he looked at Sasha. "You'll have to look after the baby. We'll get you a cart and he can sit in it while you help get baby food and diapers. That's where I'm heading so I'll help you." He took another deep breath. "Let's go!"

Once we were all out of the truck and ready to go, Daniel grabbed my suitcase out of the truck bed and then the stroller. He looked at it and then at the trees.

"I wouldn't bother fighting with this thing through those trees."

Belle waved it away and turned to head towards them but I grabbed Matty's diaper bag from underneath it. I had put Mom's purse inside of it and I wasn't going to lose it.

It was just a thin strip of trees we had to cross and I could see the huge store on the other side of it. We stayed together as we crossed over and walked across the parking lot that ran behind the store. There was a line of cars parked in a row marked staff parking and some of them had doors hanging open and hoods up like people had tried to get them started before just walking away. As we got closer to the store we heard cursing coming from under one of the car's hoods and a loud bang as the hood was slammed down. An older man stepped away from the car he had been working on and put his hands on his hips. He caught sight of us walking towards the store and shook his head before heading towards us.

The older man was still shaking his head when he met up with our group and he started talking right away.

"Sorry folks, store's closed! We lost power a few hours ago and it never came back on so I sent the staff home and just locked it up. Now it seems the cars aren't working either!"

Daniel Ross quickly explained about the bombs and what was coming. The store manager's face went from disbelief to panic before he made a move to run past us. Daniel grabbed his arm to keep him with us.

"We need the keys to the store! We need some supplies from in there."

The store manager started shaking his head. "No, I have to go get my wife. I need to go. Besides, none of the cash registers work, I can't help you!"

"Listen to me! Things are going to get really bad in a few hours when people figure out what's happening. There will be looting and fighting and your store will be hit hard. I'm trying to give these kids a chance to survive that!" Daniel pulled an envelope from inside his jacket and handed it to the man. "There's ten grand in there, take it and we'll call it square. There's no way we can carry that much stuff out of your store with no working vehicle. Help us out here and we'll lock the door on our way out and leave the keys in your car." When the man still hesitated, Daniel's face went hard. "You can either help us now or we'll wait till you're gone and bust the door open!"

The manager looked back at his store before looking down at the envelope filled with cash. He took a step back and fished a set of keys out of his pocket and threw them on the ground between us. With a glare at Daniel and the money clutched to his chest, the guy bolted past us and around the corner of the store. We all watched him go and stood, staring at the corner like we thought he would come back until Daniel finally got us moving.

"Well, that was easier than I thought it would be! Let's get going, we have a lot of work to do and the clock is ticking."

As we made our way towards the back door, Belle kept staring at Daniel with big amazed eyes until he looked at her.

"I can't believe you just gave that guy ten grand for us!"

He shrugged his shoulders and shook his head. "That money isn't worth the paper it's printed on. Money won't mean anything in the coming days. Food, water, heat and safety will be all anyone wants. That reminds me! After you get all the food and water you need to stock up on warm clothes and blankets."

Belle looked at him in confusion. "It's August and I'm sure this will all be resolved before winter hits."

Daniel came to a stop and closed his eyes in patience. "Over two thousand bombs dropped that I know of. That will send so much debris and ash into the atmosphere that it will block out the sun. It's going to get cold, really cold for a very long time."

He strode away from us at a quicker pace and quickly got the back door open. We filed into the dim back room that was filled with pallets of shrink wrapped goods and stood around not sure what to do next. Daniel locked the door we came in behind us and immediately started searching the huge area we were in. He checked every door he could find before finally coming back to our little group.

"Ok, I don't think this place has a basement but I found a staff room that should work. There are no windows and it has a lock on the door that you can re-enforce. There's also a small bathroom. Let's get out into the store and get some carts. We need to get to work."

We followed him like sheep out onto the main floor of the store and up to the front where he gave us all carts before we scattered to different departments. I was nervous about being separated from Matty, but if the guy was telling the truth about what was going to happen then I knew we would need everything we could get so I hit the food aisles and started dumping stuff into my cart.

It didn't take long for my cart to be overflowing so I headed back to the staff room with it. There were already two carts filled to the brim at the door. One was filled with water bottles and the other with baby supplies. I left my cart with the others and headed back to the front of the store for a new one.

I filled seven carts of food in under twenty-five minutes. When I dropped the last cart off at the staff room it was in a line of others that Belle and Sasha were starting to unload. The back door opened flooding the dim area with natural light, causing me to jump. I breathed out a sigh of relief when I saw it was Daniel. He must have been loading his truck with the supplies he had come for. He gave me a brisk nod and went to the carts that were waiting to be unloaded. He scanned each one and glanced into the staff room to see what the girls had unloaded before waving at me to join him out in the store.

As we walked back to get more carts he looked down at me.

"You did good with the food, kid. I'm out of here in less than twenty minutes and all the stuff I came for is packed in the truck. We're going to head over to the health section and I'll show you what to grab from there before I go into the pharmacy. After that, I'll head over to hardware and camping. I'll get you guys some camping gear and fuel for heat and to cook with as well as some padlocks to install on that door." He pulled a watch out of his pocket that still had a tag on it and handed it to me. "This is a wind-up watch so it'll still work if you keep it wound up. After I'm gone you'll only have two hours before you need to lock that door for good so grab whatever else you can think of but be inside when it hits the two-hour mark. Radiation poisoning is an ugly and agonizing way to die."

I only nodded at his warning. As mad as I was that he had not stopped to help Matty and me earlier, I had to admit that he was making up for it with all he was doing now. He could have split as soon as he had what he needed but he was staying to help us. We got to the health section and Daniel started to point out what I should take.

"All the multi-vitamins, all the C's and D's and omega's. Take all the hand sanitizer and disinfectant as well as the iodine. The iodine is very important! You and the girls need to take a teaspoon of it in the morning and at night and give the baby one teaspoon a day. It helps with the radiation. Grab a bunch of different sizes of bandages as well then go for the painkillers and cold meds. Don't forget children's medicine for your brother, he can't take the adult stuff. After your cart's full, come back with another one and start on the hygiene stuff like soap and toothpaste and toothbrushes. I'll see you back at the staff room."

He ruffled my hair as he walked passed me just like my Dad used to do making me think of him. When Daniel had said that Vancouver had been bombed I blocked out the fact that my Dad lived in the city's southern suburb. I wondered if that meant he was dead and if I would ever know.

My mind was in a fog as I swept pill bottles from the shelves into my cart. The word *orphan* kept replaying on a loop. All I had left was Matty and I knew I would do anything to protect him. Tears kept coming to my eyes but I refused to let them fall. I had to be strong now. My baby brother was counting on me.

By the time I pushed my full cart back to the staff room Daniel had already returned and was screwing the extra locks to the door. He was giving more instructions to Belle as he worked. She was holding a duffle bag open and looking into it.

"Once things settle down, communities of survivors will form. Kindness to strangers and good will towards others will be gone. Each person will have to have value in skills or goods to rate protection. It might not happen at first but as time goes by and food and water become scarce it will turn cutthroat. The pills and medications in that bag will be your security. Get some Ziploc bags and divide them up and then hide them in different places. Be very cautious because stealing will be the norm. Keep them hidden for as long as you can until the shift happens and then use them sparingly as currency. The hardest part for you will be hardening your heart to others. Your natural reaction will be to help in any way you can. I'm not saying don't help others but always remember that your help to others will be taking away from the ones you love the most." Daniel tightened the last screw on the third lock he had installed and stepped back from the door. He rubbed his chin in thought before looking around the room at all the supplies the girls had dumped everywhere.

"Alright, I'm going to hit camping supplies next and load you up but after that, I have to go. The last thing I'll do is move that pallet of boxed toasters in front of this door. I'll leave enough room for you guys to squeeze out but it should keep anyone from looking too closely behind it and finding this door. Don't forget clothes, winter gear and that the baby is going to get bigger in the future so he'll need larger sizes, especially footwear. Grab some sewing kits to help make new things as well." He took another look around the room before leaving.

Belle and I stood staring at each other for a minute. We both were thinking about the future Daniel had painted. She shook her head and sighed deeply before reaching out and rubbing my arm.

"We'll be ok. We have a really good shot here with all this stuff. Come on sweetie, let's get the rest of the stuff."

I could only nod my head as I looked over at my brother who was bouncing happily in a baby seat surrounded by attached toys that Sasha had brought from the baby section. He was giggling at a squeaky cow and I wished I could be as oblivious to what was coming as he was.

I followed Belle through the clothing and shoe departments as she piled both of our carts with warm clothing in larger sizes for Matty, Sasha and me. She kept saying how lucky we were that the stores had started bringing fall and winter wear out so early. When we made it back to the staff room, Daniel was again waiting for us. He stood outside the door with a hydraulic hand dolly. He gave us a hand dumping all the clothes into the disaster of supplies in the staff room. Once the carts had been emptied and moved away he grabbed the dolly and used it to pull the pallet of toasters over in front of the door. Belle and I both made sure we could squeeze past it when we needed to get out.

The three of us stood in the dim stock room and stared at each other for a moment before Belle stepped forward and placed a sweet kiss on Daniel's cheek.

"We can't thank you enough, Daniel. You've done a truly kind deed by helping us. I hope one day we will meet again and I hope you and your family stay safe."

Daniel stared at us for another moment before nodding.

"Good luck."

He turned to leave and took two steps before turning back.

"Stay as long as you can in that room. The longer you stay the better chance you'll have to survive in the future. Stretch it out until the last drop of water is gone!"

At our nods of understanding, he turned and disappeared out the back door. Belle and I just stood there again, lost in our thoughts, until Matty's faint shriek of happiness jolted us back to the present. I looked down at my new watch and then up at Belle.

"I think we should get more water."

Her forehead was creased with worry but she nodded her head. "I think you're right. Let's go."

Chapter Five-Skylar

As I flew across the stage on feet as light as air, the audience cried. It was my big moment. I had practiced the choreography for months for this moment and the audience was ruining it. My pirouettes were perfect and the costume I wore floated and shimmered at my every move. It was the best performance I had ever given but still the audience cried. A hand roughly shook me out of a spin and I turned with anger to see my dad's grief-ravaged face. There was so much pain in his face I had to look away and my eyes settled on the screaming baby in my arms. Reality crashed down on me as I realized I wasn't at my dance recital but sitting on the floor beside my dead mother and the fear and anger surged through me.

"Where were YOU? You left us and I couldn't..." A sob broke my voice and I couldn't go on.

His head was bowed as he clutched at Mom's hand and the sounds coming from him were so foreign for my strong dad that I skittered away from him and Mom. With the baby clutched to my chest I heaved myself up and onto the couch and just watched him fall to pieces.

I have no idea how long we sat there with the baby screaming, Dad devastated and silent tears pouring down my face. I jolted away from him like a startled animal when in one smooth move he rose to his feet with Mom in his arms and carried her out of the room. My mind was still in the fog of grief and horror as I bounced the baby and patted his back but nothing would soothe him. It was like he knew his Mother was gone and he grieved with me.

I didn't register Dad coming back until he tried to take the baby from me. The noise that came from my throat was close to a growl causing him to take a step back. He swallowed hard as he scanned my face and then held out a baby bottle to me.

"He needs to eat. That's why he won't stop crying, he's hungry."

I snatched the bottle from him and the way the baby's lips clamped onto the nipple told me he was right but I was so full of fury I snapped at him.

"He's crying because his mother is dead!"

I saw Dad flinch at my harsh words but I didn't care. This was his fault. He left us and Mom died. I turned my back to him and hunched over the baby like a lifeline. I had no words for him when I could barely think myself. I just stared at the tiny boy in my arms as he gulped down the formula in the bottle. As his little eyes fluttered down in sleep I briefly wondered where the formula had come from. There hadn't been any in the bags I snagged from the nursery.

"You have to burp him." A sad voice said from behind me. I slowly turned around to face Dad. His face looked older and grey but his eyes were glued to the baby in my arms. I wanted to throw myself at him and have his Dad arms pull me close and make everything ok again, but a glance at the blood-stained towels at his feet made me realise that nothing would ever be alright again. I stepped towards him and held the baby out.

"His name is Benjamin…she named him…before."

Dad's face crumpled in grief as he reached out and took his son. Resting the tiny body against his chest he started to gently pat him on the back until a soft burp escaped him. Dad's eyes met mine and he opened his mouth to say something but before he could I turned away and raced into my new bedroom, slamming the door as I went. I snagged the iPod and jammed the earbuds into my ears before throwing myself onto the bed. As the music crashed over me I blocked everything that had happened in the last few hours out of my mind and let myself sink into oblivion.

"Sky, Skylar? Honey, you have to come out. You have to eat something! Come on, Sky! It's been two days!"

I jammed the pillow over my head harder to block out his voice. I didn't want to see or hear him. If I went out there, I would have to start living a life without my Mom, without my home, without my life. I just wasn't ready to accept it all yet. I had sneaked out a couple of times to use the bathroom but other than that I've stayed in bed with my music and dreams of what used to be.

Hours later, a cry broke through my sleep sending me bolt upright. I fumbled around in the covers for my iPod and saw that I had killed the battery. My bladder lurched in need so I docked the device to charge and crept over to my closed door. Cracking the door a fraction of an inch I peeked out and saw no one in the main room so I raced over to the bathroom.

After I had done my business, I stopped in the kitchen to grab something to eat. The lights had been dimmed so I propped the fridge open so I could see what was in the cupboards for a snack. I didn't want to talk to Dad but he was right that I needed to eat something. I grabbed the closest thing I could reach which was a granola bar and promptly dropped it to the floor when a pitiful cry rang out.

I snagged the bar and crept back towards the bedroom but paused at the door before going in. The baby was making the saddest little whimpers that tugged at my heart. Dad's bedroom door was only halfway closed so I leaned in and gave a quick look. Dad was sleeping in the big bed and tears sprang to my eyes when I saw that he was clutching what should have been Mom's pillow. Even in sleep, his face was lined with sadness and I immediately felt shame at my behaviour. He had lost his wife, his best friend and I had punished him and left him all alone with his pain.

I swallowed hard and stepped into the room. The baby was laying on his back waving his tiny baby fists in the air. Dad had made a little bed for him in a dresser drawer. I reached down and gently ran my finger over his forehead and down his little nose and almost laughed out loud when he snatched it with his small little fingers. My eyebrows shot up at how strong his grip was and I couldn't stop the grin that spread across my face. I felt even worse about the way I had been acting when I realised that I had left my baby brother too. The thought of how Mom would be disappointed in me sent a wave of pain crashing in my chest and a small hitching sob escaped.

In one smooth motion I reached down and scooped the baby up and against my chest and half turned to leave when I caught sight of Dad looking up at me from the bed. I opened my mouth to say something but nothing came out and tears burned my eyes. He just shook his head and brought his big hand up and covered mine and half of the baby's back. After a few minutes, I just nodded at him and carried the baby out into the living room. We settled on the couch and I sang him baby songs that I remembered from when I was little. He may have lost his Mom too but he would have the best big sister any kid could ever have.

Chapter Six-Rex

The door is locked and the staff room is overflowing with supplies that we're trying to sort and store in some sort of order. Matty had fussed until we set him free from his baby seat and now he's trying to climb every pile in the place. I feel like I'm spending more time pulling him away from stuff than getting anything put away so I give up and just grab him and fly him over my head like an airplane. His shrieks of happiness soothe the burn in my chest that never seems to go away completely. I keep waiting for the cops to show up and arrest us for looting the store but the hours drag on and no one comes. I'm also waiting to feel sick. I don't know anything about radiation or how it works or what it does to a person, but Daniel said it was a really bad way to die. The more time that goes by and I feel fine, the more I wonder if maybe he was a crackhead that didn't know what he was talking about. Or maybe he knows what he's talking about and we're safe in this concrete room. I guess there's no way to know.

"Rex, watch out for those diaper boxes!" Belle calls out as my foot catches the corner of one and I almost drop Matty. The place is a total mess with hardly any clear floor space. I don't know how we're going to be able to stay in here for a couple weeks without going crazy. I bounce Matty on my hip as I look around the room. Grey painted cinderblock walls with posters reminding employees to wash their hands stare back at me. There's a scratched up fake leather couch against one wall currently piled high with winter clothes. A short kitchen counter with a small sink and two microwaves stacked on it has some cupboards that we can fill up and a staff room table that seats six. The other wall is filled from floor to ceiling with staff lockers and is interrupted by a door that leads to a tiny bathroom. The whole room is maybe twenty by twenty and with the four of us and all the supplies we've gathered and dumped into it, I can't even think of how we will be able to stay here for so long.

This is stupid! We need to get out of here and see what's going on out in the world. I'm just about to tell Belle and Sasha how that Daniel guy was probably a crackpot when yelling and a crash comes from out in the store. We all freeze in place as the sounds of people get louder and it doesn't sound like happy shoppers out there.

Belle dashes over to the table and reaches for the lanterns that light the room. She's waving frantically at me to get to the other lantern on the counter as she turns the flame down low so the room is hardly lit at all. I do what she wants but I don't really know why until she comes over and scoops Matty away from me. She digs a handful of goldfish crackers out of an open bag and transfers them to Matty's sticky hands as she whispers to me.

"The light might show around the cracks of the door. We don't want anyone to know we're in here!"

I glance at the door for a second and then back at her.

"Do you really think that guy knew what he was talking about? I mean how would he even know if all that stuff was true?" Maybe one of us should go out and check things out."

Belle is shaking her head even before I can finish talking. She opens her mouth to reply but before she could say anything there is a scream of anger much closer. The sound of a woman's voice coming through the door is shrill as she yells.

"You can't keep it all! We need things for OUR families too!"

We all have our eyes glued to the door in shock. Her voice sounds like she is right outside the staff room. There's another crash and a bang of a door hitting a wall before a pain filled wail erupts from the woman. I turned back to Belle in scared confusion as the wailing moves away. Belle's face is filled with terror and her eyes are huge as she lifts a shaking hand and puts a finger to her lips in a "shhh" gesture. I slowly nod my head and glance over to Sasha. The poor girl is down on the floor between piles of supplies. She's got her knees drawn up and her head down on them as she rocks back and forth in fear. I almost jump out of my shoes when a deep male voice speaks next.

"We're going to leave everything back here for now and just start loading the stuff from the front up. Get a few of the men and go to the other doors to the storeroom and barricade them. Move some pallets in front of them so they can't be opened from the main store then get back over here and we'll topple some of the shelves from the front of the store. I don't want anyone back here! We're going to need the stuff back here in the future so block it off and we'll come back in a few days for it."

There's a muffled reply to these instructions then what sounds like the pounding of boots on cement as people run away. We're all just frozen in place as we wait to hear what will happen next when Matty starts to squirm. Belle tries to hand him more crackers but he just shoves them away and squirms harder. Belle's face is cracking in panic and even Sasha has stopped rocking and looking at us with a scared expression. I look around the crowded room for something; anything to distract him from the Matty bellowing that's on the way and spot his Sippy cup. I grab it and snatch the kid from Belle making a beeline for the tiny bathroom. I gently close the door and slide down against it to the floor with Matty in my lap. I raise my knees up and lean his back against them before handing him his cup. His small feet push against my chest until I start rubbing his tummy. This trick always works to settle the kid down. Mom used to beg me to use it on him when she was at her wit's end. It only takes a minute for his little green eyes to start fluttering closed. It's been a crazy day and the kid has held up great but right now silence is life or death so I hold my breath and silently beg him to go to sleep.

My breath whooshes out and Matty's eyes pop open with a squawk when there's a huge crash out in the main store room. It sounds like metal against metal and I feel the crash through the wall. Matty's face is screwing up into a rage so I grab a tiny foot and blow a raspberry against it. It's enough to get his attention. He pushes it against my face for another with a small smile pulling at his lips so I do it again and then start back up with rubbing his belly. I don't know how long I sit there but Matty falls asleep at some point and I let the tears that have been stuck in my throat for hours just trail down my face. Any hope that the world didn't suck as bad as Daniel had said it would is now gone and that means he was probably right about everything. My mom is gone and now I believe that Dad is dead too and that means me and Matty are orphans.

Chapter Seven-Skylar

"Is that a cow?"

I rub my weary eyes after a night of staying up with the baby. I'm trying to figure out how this sling works that I found in one of the bags I had brought from the nursery. It didn't take me too long to figure out that it's next to impossible to do anything with a baby in my arms. The problem is, the sling is way too big for my skinny ten-year-old body so I finally have to tie a knot in the fabric to make it small enough that the baby stays pressed against me. Thank God there was a little paper handout in the package or I would have been totally lost on how to use the thing.

Dad had passed through the living room earlier and had left the door to the rest of the cave open so the sound was clear but it didn't make sense. As soon as the baby was in the sling I went through the door looking for him. I had only had a quick look through the door on the first day when we got here so I was unsure of all that was back there. There were a few doors that were closed to rooms that had walls but the walls didn't go all the way to the top of the rock ceiling. Looking up I could see lights hanging from a grid of metal beams drilled into the rock and it amazed me that Dad had created this place.

Past the closed doors, there were the shipping containers to the left and to the right was an empty gardening center. I counted six raised rows of empty beds filled with soil with hanging grates above them that held a lot of empty pots waiting to be filled. Past those on the far right wall were two more closed doors that were embedded into the rock wall. The size of this place was beyond anything I had imagined when my parents had talked about Dad's man cave. The distinctive sound of a cow came again and I picked up my pace further into the cave.

Once I had passed the shipping containers and the garden area I again came to a stop at the sight of the weirdest thing. A knee-high wooden fence cut all the way across the cavern and on the other side of it was a four foot across stream. I tracked it across the room and saw a small pond cut into the rock floor. Looking closer at the water I saw a flash of silver as fish shot by. A silly grin split my face. Not only did Dad make a crazy huge man cave but he put in his own fishing pond. There was a little, railed bridge to cross over it so I crossed and went deeper into the back.

To the right were more wooden fences but larger and made into two animal pens. One of them had a chicken coop and my eyes popped out at the six hens and one ugly rooster that were scattered around it pecking at the hay covered floor. The other pen had a real honest to goodness cow in it. It was huge and black and white and it stared back at me over the rail as it chewed hay. I stared back in amazement. Where on earth had these animals come from and more importantly, was I going to have to clean up after them?!

The scream of an electric saw had me turning to the left and there was Dad. My eyes roamed over the walls and tables around him that had all the workshop tools anyone could ever need. I had to shake my head a little in disbelief, Mom was always going on about how many tools Dad had in the garage but this was at least double that.

I wandered over towards him and saw that he was in the middle of a building project. From what I could see it looked like he was building a baby cradle. I just stood there watching him and rubbing the baby's soft head while he worked. His face was haggard and strained and he looked like an older tired version of my dad. Again, sorrow at the loss of Mom rushed through me. Tears burned my eyes as I tried to deal with how unfair this change in our lives was.

Without looking up at me, Dad spoke.

"Can't have the boy sleeping in a drawer, you'll be surprised at how fast he'll grow out of it."

I glanced down at the tiny lump against my chest for a second before clearing my throat.

"Um, Dad? How did you do all of this? I mean all this stuff, the lights and the rooms and the containers, where did it all come from?"

It took him a few seconds before he answered me as he continued to sand the board of wood he was working on. He finally dropped the sandpaper and looked up.

"It wasn't me. I didn't do all of it. I just did the finishing work, our living area, and the animal pens and gardens. Uncle Bill had the majority of all this done before I even knew about it."

I shook my head, "He just gave all this to us?" I asked in disbelief.

Dad ran a hand through his messy hair making it even messier while looking around the cavern.

"Not exactly, it's complicated." He paused and frowned before shaking his head. "I guess it doesn't really matter anymore but what he did was very illegal. After I got back from my last deployment your Uncle and I spent a weekend camping out up in these mountains. We talked a lot about the future and the way the world was headed. I told him about my plans to find a piece of land up here that I could buy and set up as a fallback position in case the sh… uh, poop hit the fan. We talked plans and designs and supplies. It was sort of fun to lay it all out. After that, I got busy and spent the next four years building up the construction company. There just never seemed to be any time to dedicate to working on the plan and I just tucked it away for someday in the future. Do you remember when Uncle Bill and his family came out to visit for your sixth birthday party?"

I nod my head slowly with my mouth flattened into a grim line, I remember that visit very clearly mainly because of the most annoying boy on earth was with them. Uncle Bill's son, Jackson is a few years older than me and a total spaz. He had ruined my party by popping the heads off and plucking the wings from the two Fairytopia Barbie dolls I had gotten as gifts. This had enraged me so much that I had smacked him in the head with the piñata stick causing him to fall backwards onto my Barbie birthday cake. Party…Ruined!

A brief grin flickered across Dad's face at the memory. "Right, um do you remember after the party Bill and I went for a drive? Well, he brought me here. I was blindsided! I couldn't believe what he had created; I mean the scope of the project was just huge! Apparently, he had started working on it right after that camping trip. No one official knew about it. It was totally off the books and he had used foreign contractors to get it all done. He told me about the twin shelter back east and how he had massaged the figures to fund building this one. I've been working on the finishing touches for the last four years. Uncle Bill still has a hand in some of it. Every now and then I would get a text from him that I should go enjoy the mountains and that's when I knew a delivery was coming. All these shipping containers were his doing. So he's the reason we're here. He saved our lives."

Tears immediately flood my eyes as I think about Mom. She's not here, she wasn't saved. Dad's face crumples for a split second before firming up again so I angrily swipe at my eyes and clear my throat. If I let the tears come I know I'll be back in my room and this time I might not come out.

"So, Uncle Bill made all this for just us?"

Dad shakes his head, "No, it's for a lot more than just us. We're sort of caretakers for it. Bill knew that if something big happened that our country would need help afterwards to rebuild. The shelter in the east won't be enough. Most of our population is in the east so it makes sense that they would start there but this is a huge country and there's a lot of resources that will be needed from the west. Unfortunately, our politicians seem to mainly focus on only two of the provinces. Bill wanted to…diversify."

I didn't really understand everything Dad said but I got the gist of it. I looked around the cavern again before asking,

"Seems a little small to be used to rebuild doesn't it?"

Dad snorted out a laugh. "Oh Sunshine, this is just the tip of it! There's two more levels underneath us filled with materials and supplies as well as a hydroelectric plant even deeper to provide endless power. We would be able to clear the area outside and provide everything to start rebuilding when the time comes."

My mouth had dropped open and my eyes were huge as I listened to his explanation.

"That's amazing! When are we going to go get the people? Isn't it dangerous for them to still be outside? What about kids? Will there be a school set up?"

My excitement and enthusiasm dimmed as he was shaking his head and then disappeared and turned to shocked anger at his words.

"No Sky, no one can know about this place! It would be overrun in minutes! No, Bill has a plan in place, a timeline for how to rebuild and when and it won't be for a long time. We just have to wait it out."

I swear I felt my heart harden as he spoke. Here it was again, my dad looking away instead of helping people. My tone was harsh and sarcastic for a ten-year-old but I didn't feel like a kid anymore.

"Kinda hard to rebuild if everyone is dead, Dad!"

His face flushed and he snapped back, "You don't understand, you're too young to understand what would happen if we tried to help!"

I glared at him and delivered the most painful, mean thing I could.

"Oh, you mean like driving past those kids on the road? That was to save ourselves, right? Hey, that sure worked out didn't? Oh wait, no it didn't, cuz Mom DIED anyway!"

My voice cracked on a sob on the last word so I just spun around and ran back to my room with the baby wailing against my chest. Dad might have saved our lives but what kind of life would it be knowing that he let all those people die?

Chapter Eight-Rex

It felt like the screaming and yelling went on forever. There were even a few gun shots that scared the heck out of me. The only good thing was that they were far away and no sound was coming from the storeroom. My legs are numb and fill with pins and needles as I try to get to my feet without waking Matty up. As much as I want to stay hidden in this tiny bathroom, it's kinda gross and I should be helping Belle and Sasha with the organizing.

Matty's good and out, a dead weight in my arms with a line of drool down his chin so I just hoist us up and open the door. Things seem to be settling down out in the store, at least I haven't heard anything from out there for a while so maybe they've taken what they wanted and left. I'm scared just thinking about what that guy said about coming back for the stuff in the store room. I don't know what he'd do to us if he found us. Nothing I can do about it right now so I look around for the girls but all I see are piles of stuff stacked everywhere. My belly clenches and my heart stutters for a second when I think that they left me and Matty here, but then I see the top of Belle's head against the back wall and the breath I didn't know I was holding comes whooshing out.

It's kinda crazy. I mean, I've only known the girls for a few hours but the thought of them leaving me terrifies me more than the gunshots out in the store had. I don't think I can do this on my own. I don't think Matty and I could survive without them. I push that thought down and away as I head over to them. Belle's sitting on the couch with Sasha laying down on it with her head on Belle's lap. Looks like Sasha's sleeping just like Matty.

Belle sees me over the stack of supplies and puts a finger to her lips then gently slides out from under her daughter and stands. I can't help but be jealous for a second. I wouldn't mind putting my head down on Mom's lap and drifting away from all this madness. My face forms up into a scowl. No point thinking that way now. Mom's gone and I have to suck it up and be a man for Matty's sake.

Belle shoots me a questioning look at my angry expression as she reaches for Matty but I just look away with a shake of my head, can't talk about it right now. I see there's a small area that's been cleared and they've set up a playpen with soft baby blankets in it. I feel my shoulders slump in relief that Matty will have a safe place to sleep. I don't know where the rest of us will be sleeping but at least the kid will be fine.

Belle gets him settled in and covered up and then leads me by the elbow away from the sleeping kids. When we're far enough away that we won't wake them she turns me to her and tries to give me a hug but I can't right now. I can't be sad and I can't be soft so I push her arms away gently and shake my head again. She's got the softest blue eyes that are so filled with kindness that I have to look away from them too.

She lets out this sad sigh before telling me, "It's ok honey. I'm here when you're ready. I promise I'm not going anywhere."

I press my lips together tightly so I won't yell at her that she can't promise that. I know she's just trying to help. After a few seconds, she just gives my head a quick rub and goes on.

"So, I don't think there's anything we can do about those men that were outside talking. If they come back, we'll just have to keep very quiet and hope they don't find the door. So let's just worry about that when or if it happens, ok? Now, I think we can get a lot of this stuff out of the way by filling up the lockers and some of it can go into the bathroom and the kitchen cupboard. After that, we'll just stack the rest along the walls. Once we've made some room we can blow up the air mattresses that Daniel brought back from the camping section."

I'm nodding along with her but then when she mentions Daniel I have to ask her.

"Do you think that guy was telling the truth…about what happened out there? You know the bombs and stuff?"

She looks at me thoughtfully for a second before she starts nodding her head.

"Yes, I do now. At first, I was just caught up in getting the supplies but then when nothing happened I started to doubt what he'd told us. After all those people came into the store and all the yelling and gunshots, well, if something hadn't happened then where were the police? Stores getting looted and gunfire just doesn't happen here without the police showing up. So, yes, I think somehow Daniel Ross knew what had happened and he tried to help us so I'm going to do exactly what he told us to do."

I tip my head in a sharp nod of agreement. There's nothing else I can think of to do so we'll follow his plan for now and see what happens.

"Alright, let's get this stuff sorted out then."

Belle and I move and organise as quietly as we can but eventually Matty wakes up with a squawk and that wakes Sasha up, so we put together a meal of soup and sandwiches, cooking them on the camping stove. Then back at it until there's finally a decent amount of floor space to blow up the mattresses. I remember Dad doing this before and it was a quick flick of a switch and a small pump blew air in. It only took a few minutes to inflate them but now flicking a switch does nothing, so my leg's ready to fall off after using a foot pump to get the first mattress inflated. I flop down on it in exhaustion while Belle pumps up the second one.

The store's quiet and I hope that means we're alone in here. I'm bushed and I don't want to have to deal with anything else tonight. Too bad that's not going to happen. Matty crawls on top of me and starts slapping at my cheeks. The kid's bored and I'm his favorite toy. I close my eyes for a second and just endure his patty cake on my face when the smell of a toxic Matty bomb hits my nose. Oh man, this kid is a master of gross! I do a quick peek at Belle to see if she'll sweep in for the rescue again but she's pumping away at her mattress and I can't help but frown at how tired *she* looks. I'm on my own for this one.

I roll the kid off me onto the other side of the mattress and he giggles when he bounces. I stagger to my feet and grab the diaper bag. There's a folded up plastic changing mat that Mom always uses when we're out somewhere so I get that laid out and snag the wiggle monster onto it. Even with kicking legs and arched back I manage to get his pants off, letting out even more stink. I take a quick look around hoping Daniel had somehow managed to get us a hazmat suit but no luck. I do spot a package of painter's dust masks. I have no idea why they are here but I grab it and tear one out and get it on my face and nothing, still reeks, but the kid has frozen in place as he stares at me. I'm about to pull it off, afraid he's going to start wailing, but the biggest grin splits his face and he starts to belly laugh the way only Matty can. Good enough for me. If it keeps him happy I might be able to get him clean without poop flying everywhere.

Not only do I get him decontaminated, but I also give his small body a wipe down with the baby wipes and into some pajamas that are in the diaper bag. I know Mom always gives him a bath before bed, but water's too important now to do that so hopefully the wipes will be good enough. I lean back, happy to have got the job done only Matty lets out a screech and reaches for me. I pull off the painter's mask and hand it to him. That's all it takes to make the kid happy again. He spends the next ten minutes putting it on his face and pulling it off while yelling BOO before belly laughing again. I leave him on the mattress playing while I go into the bathroom and do my own clean up with the wipes. I use a tiny bit of water from a bottle to brush my teeth while trying not to yawn. I'm so tired, I just want to crash. A quick thought passes through my head that this must have been how Mom felt at the end of a busy day dealing with me and Matty. I shake it away, still not ready to deal.

By the time I come out of the bathroom, Belle's done with her mattress and has blankets on both. Sasha has Matty on the couch with a bottle and his eyes are starting to droop closed. I go over and gently pry his little fingers from the painters mask. I shoot a quick look at Sasha's face but she seems almost as zoned out as the kid so I start humming the song Mom always sings to him at bedtime. His lips curve up around the bottle but his eyes stay closed so I keep humming even though the tears I said I couldn't deal with are pouring down my face. I keep humming as pictures flash through my mind of Mom and all the cool things she used to do with me, from beating me constantly at Halo to tackling me during laser tag. The hateful words I said to her over the last month to the way she looked after the van crashed, they all keep circling around in my head until my humming turns to sobs and then Belle's pulling me away. She's holding me in her arms and rocking me and I finally just let it all go.

Chapter Nine-Skylar

Here's the thing about living in a cave, there's nowhere to go. I'm mad, I'm mad at my Dad, I'm mad at Mom that she's gone and I'm even a little mad at the baby because he needs me so much. I just want to run away and be alone and mad but I'm stuck here and the stupid computer say's it'll be years before it'll be safe to live outside. I heard Dad talking to it about forecasts and it says something called nuclear winter is coming. It's August so I don't get how any kind of winter will be here for months but who cares, it just means I'm stuck in here.

The days just seem to blend into each other with nothing changing. I feel like I'm in a fog and I keep waiting for something, anything to happen that will change our situation. I can't get out of the shelter so instead I explore it. With baby Benny strapped into his sling, I roam through the different levels and areas. I go through every door that I can open and kick the ones I can't in frustration. I'm beginning to hate the sound of AIRIA's voice when she says,

"Skylar Ross, unauthorised entry."

I've made so many laps through this place which I'm beginning to see as a prison, that by the end of the day my feet and legs ache.

I'm bored and I just can't imagine being stuck in here for the rest of my life. There has to be more. As my frustration and anger grow day by day there's only one outlet and that's taking it out on Dad. He always seems to be busy with building something, taking care of the animals or planting seeds in the grow area. My attitude and sarcastic tone seem to bounce right off of him which makes me even madder. I'll admit it, I was a huge brat! So it almost comes as a relief when he finally loses his patience with me.

Ten days, it's been ten days since my life ended. I lie in bed staring up at the ceiling and wonder if I should even bother getting up. I mean, what's the point? Another day of just wandering around my prison seems so pointless so I roll over and close my eyes.

"Skylar, get your butt out here!"

My eyes pop back open. This is new. Dad hasn't really talked to me in days. I sort of want to ignore him and just stay in bed but I worry it's something to do with the baby so I roll out and stuff my feet in slippers before shuffling out into the living room. I head straight over to the cradle Dad made to check on the baby but Benny's sleeping soundly so it's not him. Then I see something different, piled on the dining room table we haven't even used yet is books, papers, pens, pencils and DVDs. I wander over and see that there's a plate with an omelette and toast waiting for me as well as a folded sheet of paper leaning up against a glass of orange juice with my name on it.

I glance over at Dad but he's got his back to me working at the stove so I snatch the paper up and unfold it. It's titled Skylar's Schedule and I frown as I read. Eight to twelve, schoolwork with each hour broken down by subject then lunch. One to three is marked Agricultural Studies then three to four special interests. I have no idea what that means so moving on is dinner, family time and hobbies from five to nine and then bed. I scowl over my shoulder at his back, nice of him to give me one free hour a day to do what I want. I think about tearing the schedule to pieces before going and slamming my door, but at the same time, I think it might be good to start doing more. I can't just wander aimlessly and I **had** just been thinking that there needed to be more to life, so maybe I could try this plan out for a while and see how it goes. BUT, but no cow poop shoveling!

The truth is it felt good to stretch my brain with school work. It was a bit of an escape to not have to think of our situation and just concentrate on equations and literature. Agricultural Studies turned out to be learning about the cow and chickens and how to care for them and keep them healthy. It also meant learning about growing things. There was definitely some EWWWW moments but it was kind of fun.

Dad didn't know a whole lot about farm animals and crops so together we learned what we needed to know. And our teacher, the annoying AIRIA! We watched videos and tutorials on everything from how to make baby chickens to crop rotation. We learned how to can the things we were growing and how to make butter and cheese. It was interesting and different and we did it together with Benny strapped to one of us the whole time. Slowly we learned how to be a family without Mom.

Special Interests, AKA how to be crafty. He tried to teach me wood working but the glazed look in my eyes must have discouraged him, so he moved us on to learning how to make clothes - which was way more fun for me than him I'm sure. After I had mastered the sewing machine I tried my hand at knitting. I loved knitting, but having so much time on my hands can mess with a person's head and I got a little carried away. Benny now had enough knitted baby clothes to wear a different outfit twice a day for a month. Dad finally called a halt on my obsession when I presented him with a full man-sized bodysuit that included booties and a hood all made from bright red yarn.

Six months passed faster than I thought it could and Benny was getting bigger by the minute. Dad had made him a wooden walker with wheels and it was a relief to be able to ditch the sling. He seemed to be happy to push himself along in it while we went about our chores. I missed Mom and ached for our old life every day but it was working. There were still times when it all got to me and one of those times was when I had a major meltdown on my eleventh birthday.

The day started out great with a candle stuck in a breakfast cinnamon bun. Dad sang Happy Birthday to me with Benny clapping his hands and screeching along with him. I loved the leather journal and other treats they gave me but the main present Dad had for me sent me into a tailspin of anger and sadness. He had been working on finishing some of the rooms in the back cavern for quite a while but I hadn't really been paying attention to what he had been doing so it was a total surprise when he led me to a door that I knew was a room filled with workout equipment. The floor was no longer natural rock but a highly shined and polished wood floor. The walls had been transformed from unfinished drywall to mirrors and running along two sides of the room was a dancer's bar. The weights, treadmill and bike had all been set up properly in one corner with a punching bag hanging to the side.

I gave Dad an uncertain smile, not sure what the gift was and he just laughed. He walked over to a huge wall mounted TV and stereo and picked up a stack of DVDs that had been wrapped with a ribbon and handed them to me. I stared blankly at each DVD as I shuffled through them.

"I know how much you miss our old world so I set this room up like the dance studio you used to go to. You can watch the videos to keep learning and AIRIA has more advanced training vids in her databanks. You can keep learning ballet and any other style you want! Happy Birthday, sunshine!"

I gritted my teeth and forced out one word.

"Why?"

"What do you mean, why? You love to dance! That's all you ever wanted to do before. You wanted to be a professional dancer!"

His face was filled with confusion and all it did was make me angrier. How could he think that I would want to keep training to be a dancer? I let the DVDs fall from my hands to the floor.

"Dance is DEAD just like everything else in the world!"

I spun on my feet and rushed out of the room. It was many years later before I ever went back in.

He let me sulk and pout for a few days before he dragged me out of it. I was flopped out on my belly on my bed with the iPod blasting in my ears when he strode into my room and pulled the plug. I flipped over in outrage until I saw he had on his "I mean business" face.

“Listen Sky, I know this has been hard on you and I understand why you wouldn’t want to keep up with your dancing. I’m sorry. I’m sorry for everything that’s happened and I’m sorry you had to let go of your dream but this isn’t going to make it better. You have to channel your sadness, bitterness and anger into something else or it will eat you alive. Trust me, I know what I’m talking about. After some of the things I saw on deployment, well, it’ll eat away at you if you let it.

So, dance is dead. Let’s try something else. Get up, wash your face and meet me in the back. I think I have something that’ll help a bit.”

He walked out without another word. I lay on my back staring up at the ceiling in misery. I couldn’t think of anything that would make me feel better except for everything going back the way it used to be, but after a few minutes curiosity got the better of me so I rolled off the bed and followed him to the back rooms.

I saw him at the far end of the cavern before he disappeared through a door. I had been in and out of all the rooms of the cavern and almost all of them were empty shells, but after the work he did on the workout room I guessed this would be new too. I peeked around the door and sure enough, he had renovated this room too. It now had three lanes divided by raw plywood with a waist high counter nearest the door. The far end had stacked bales of hay with what looked like sandbags behind them. Three paper targets were attached to the hay. One was a traditional bull’s-eye target and the other two were human silhouettes. I edged in further and saw the counter was full of weapons. Handguns, rifles, a crossbow and archery bow were laid out as well as boxes of bullets and different types of arrows. He spoke without turning.

“I can’t give you back the world we lost but I can give you the skills to survive in this new world. It’ll take time and a lot of practice. You won’t be able to handle most of these yet but as you grow and get bigger and stronger you will be able to handle any of them." He finally turned to face me and raised his eyebrows in question.

I walked up to the counter and let my eyes roam over each weapon before looking up at him and nodding. A sad smile tugged at his lips.

"Let's begin. The first thing you will learn is safety and maintenance. Then we'll talk about a workout routine to help you get stronger."

I cocked my head at the weapons and asked,

"And then I can shoot a gun?"

His laugh soothed my heart and made me realize that I couldn't remember the last time he had laughed but his words had me frowning in disappointment.

"Sure Sky, we'll talk about you shooting in a few months if I think you're ready. In the meantime, you will learn and work with the bows. Deal?"

I moved over to the bows and ran my hand over the smooth arch. There might not be any room for dancing in this new world but survival might depend on these weapons so I turned to him and gave a firm nod.

"Deal!"

Chapter Ten-Rex

I don't know how much longer I can stay in this room. We started making slash marks on the wall to mark the days and we're up to sixteen. Sixteen days without opening that door. I swear we all spend time just staring at it. Being locked in here together has made us into a new family and like every family, we've started to drive each other crazy. Sasha and I have been snappy with each other like only siblings can be and Belle is frustrated with us both. Thankfully, Matty hasn't been too hard to handle except for a few times he's cried for Mom.

I'm pretty sure the room must stink but I guess I've gotten used to it. There's a pile of double bagged garbage bags in one corner that's filled with dirty diapers and empty food cans and wrappers. There's another pile in the bathroom under the sink filled with even worse. The toilet won't flush so we have to use a mop bucket lined with plastic bags to do our business. I hate the smell of baby powder now. Belle uses it on all of our hair to soak up the grease and then brushes it out. The baby wipes we stocked up on and use to wash our bodies has the same smell. It totally makes me want to gag. The smell of that is almost as bad as the garbage pile.

I used to love playing board games like Monopoly and Risk bust we've played so many games to pass the time that it's more like work now. I know I should be grateful that we're in here and safe but it's hard to feel that way when I don't know what's happening outside. Except for that first night when we heard the looting we haven't heard anyone else in the store. Maybe they picked the store clean so no one's come back or everything is fine out there. It's still hard to believe what Daniel said is true without seeing what's out there but I keep thinking the reason no one's come back into the storeroom is because they're all dead.

Matty's napping, Sasha's reading a book that she already read twice and Belle's doing inventory…again. I'm lying on one of the air mattresses watching her scribble in a journal. She's been doing this twice a day since the second day. To me, it looked like there was tons of food and water but the stacks have really gone down even though we don't ever fill up. Breakfast is always the same small bowl of oatmeal with dried fruit and a few crackers smeared with peanut butter. Lunch is a handful of nuts and a granola bar and every other day some canned vegetables. I'm always hungry during the day but Belle says we have to ration. Dinner is everyone's favorite cuz it's the biggest meal. Belle says it's so we go to bed with a full stomach. It's almost always pasta or rice with canned meat and vegetables mixed in.

Belle closes her journal with a sigh and I ask her the same question I do every time.

"How long?"

She doesn't answer me at first but looks at the door. Then she turns to me and puts on a smile that I know is fake.

"We can stretch the food for ten maybe twelve more days but the water will be out in four."

I look over at sleeping Matty then up at Sasha on the couch before I stare at the door then back to her.

"So we leave in four days?"

Four more days then we get to see what really happened out there. I'm half scared and half excited but it bumps up the scared part at Belle's next words.

"No, we can't wait until the water's gone. We have to start looking for more while we still have some or we'll get too weak." She bites her lip and looks away from me. "I'll go today."

For a half-second, I'm filled with relief that she'll go find more, but then Matty mumbles in his sleep and I get that she can't be the one to go.

"You can't go. It has to be me. If things are really bad out there like Daniel said and something happens to you, then me and Sasha would be on our own to take care of the baby. He needs you. It has to be me."

Belle looks over at her daughter then Matty before meeting my look with sad eyes and she just nods. I look at the door again and my stomach does a flip with nerves. I take a deep breath and try and psyche myself up but it's not happening so I just scramble to my feet and take a step towards the door.

"Whoa, whoa! Wait, Rex. We need to talk about a plan first and you need to take some iodine before you leave the room!"

A plan, yeah ok, that's a good idea. I already took a spoonful of that nastiness this morning but Belle says I need some more so I fake a gag to try and make her laugh.

I plug my nose to try and not taste the stuff but it doesn't really help. Belle gives me a juice box which is awesomeness because she hardly ever lets us have one.

"Rule number one, you DO NOT leave the store. This is just a scouting mission. I don't even know if you can get into the main store after those guys blocked off the doors. I want you to carefully and QUIETLY take a look around the storage room and check the doors to see if you can get out. Rule number two, DO NOT open any exterior doors. We're more protected inside from radiation so keep those doors closed. I have no idea if it lasts this long but better safe than sorry. Rule number three, if you see anyone else out there, HIDE. If Daniel was right, we have to keep the supplies we have left hidden. No one can know about or find this room. Are we clear on the rules?"

I swallow hard to clear my throat. I never thought about seeing someone else or what that could mean. I look over at the door again but quickly look away. The excitement I felt is now pretty much gone and I'm totally scared. It's been so quiet out there that I had stopped worrying about anyone coming back and finding us. Doesn't matter though, I'm going. We need water and it's up to me to find it.

Belle makes me put on like three layers of clothes and a man's winter jacket. It helps the fear a bit cuz I'm getting annoyed at the whole process and by the time I'm dressed I can barely bend my elbows or knees. She's fussing over me like a mom and debating if I should wear a ski mask or goggles when I just turn and stomp over to the door. She follows me over and helps me move the table we had pushed up against it before pulling me into a tight hug. When she pulls back we just look at each other for a second and then nod. She flips all the locks and uses the keys on the padlocks Daniel screwed on to it. Without looking back at me she turns the knob and pulls the door open.

Pitch black greets us from the other side and we just stare into it for a few seconds before Belle gasps. Cold, frigid air washes over us and Belle sort of moans and quickly shuts the door, flips one of the locks and puts her back against the door. She's shivering and her mouth is trembling while her eyes well up with tears.

"He was right. Oh God, I had hoped it wasn't really happening but he was right."

I stand there feeling dumb and confused. When we had come into this room sixteen days ago it was hot and sticky outside in the last days of August. The air that comes in is more like the coldest January day.

Sasha speaks behind me making me jump a bit. I hadn't even known she was behind us when we opened the door.

"What was that? How come it's so cold out there? Why didn't we feel it in here?"

Belle wipes away her tears before straightening away from the door.

"The four of us plus the candles and lanterns keep it warm in here. It does get chilly at night when we shut them off but not that cold. Daniel was right. The only reason it would be that cold out there in the middle of September is if we're in a nuclear winter."

Sasha stares at her Mom not understanding.

"For how long?"

Belle goes to her and wraps her arms around her and whispers, "I don't know but I think it will be for a long time."

She kisses Sasha on the head and then turns to me.

"Ski mask!"

After I cover my face with a ski mask and put a toque on my head, I grab one of the lanterns and go out the door, quickly closing it behind me. There's hardly any room outside the door, I forgot that Daniel had moved a pallet of toasters in front of it. I move sideways out from behind it into the store room and take a look around. Nothing really looks different from the last time I was out here, just pallets of shrink-wrapped boxes, so I head towards the doors we had used to get into the store. They're totally blocked on this side with more pallets of boxes right up against the doors so I can't even wiggle behind them.

Holding the lantern high to see I start walking in the opposite direction. We hadn't come down this way that first day so I scan the pallets as I go thinking maybe I'd get lucky and find water right away. So far all I see is house stuff, toys and electronics stacked up but as I walk further a disgusting smell starts getting stronger. It's not too bad yet but it's definitely getting stronger. I hold the lantern up higher and move to the right to see the pallets better when my foot connects with something and there's a metal clanging noise. I freeze and hold my breath waiting for someone to react. When nothing happens I lower the lantern closer to the floor to see what I hit. My eyes scan over the floor and a pile of empty food cans until they see a pair of boots. My heart's pounding and my mouth is desert dry as I move the light in the boots direction and find myself staring into frosted over brown eyes.

I take a quick step back and my foot comes down on an empty can making me slip backwards right down on my butt. There's a scream clawing for a way out of my throat but I can't seem to get a breath to release it and by the time I do I've seen that the guy isn't coming after me cuz he's dead…frozen solid dead. Amazingly, I didn't drop the lantern so I raise it up a bit and take a good look. I've only seen one dead person before and it was Dad's granny. She was seriously old but at the funeral parlour she looked like she was going out on the town with lots of makeup on. This guy looks like he was sick, like really sick. There are bald patches on his head and red marks and scabs all over his face. I feel kinda bad for him because he still has a half filled can of peas in his hand. That would seriously suck to be a last dinner.

The shock is wearing off and my heart's settling down so I take a better look around the guy and a smile cracks across my face. He's leaning against a torn open pallet filled with stacked flats of canned vegetables and beside it is another opened pallet with boxes of hamburger helper. I push myself up onto my feet and walk further down the row. I can't help but think about that voice from the first night saying they'd come back for all the food back here. Looks like it's ours now.

The further I walk the more food I find and I remember that the grocery part of the store was down at this end. I finally find the source of the smell, rotting produce that is now frozen. I shake my head at all that food gone to waste and poke around a bit until I see bags of apples that still look good. They're frozen but not rotten so I pull a few bags from a box and leave them out to take back to the staff room.

There are boxes piled up in front of the two swinging doors that go out into the main store but they only go half way up and there's dim light coming from the broken out windows in the doors. I search around and find some empty milk crates that I can stack and climb so I can look out over the boxes. They wobble a bit and I just know I'm going to crash down on the concrete floor and crack my head open but they steady once I lean against the boxes. Dim lighting from the store's skylights show me a mass of twisted shelving piled up on the other side of the door but there's one path that gives me an opening just big enough to crawl through.

I gotta think about this for a sec. I know I can make it through the hole and I figure that's how the frozen guy got in but then what? If anyone's out in the store I won't have a very quick escape route. So I just stand there and listen for what feels like twenty minutes but is probably only three. I think maybe I should go back and tell Belle what I've found so far but now that I'm out of the room and nothing's happened I don't want to go back. Curiosity tugs at me so I shrug my shoulders and climb. There's no glass in the door's window, just a thin piece of floppy plastic hangs from it so I don't need to worry about slicing my belly and having my guts fall out as I wiggle through and out onto a metal shelf. I look over the shelf's edge and there's more junk piled up against it so I scoot out as far as I can and try to climb over it and down to the floor. My climb turns into a roll as some of the stuff shifts and I end up at the bottom quicker than I planned but only bruised a bit thanks to all the layers Belle made me wear.

My fall also made a huge racket so I scramble on hands and knees over to an empty table that has price stickers for bakery goods and slide under it. I crouch in the darkness as my nose twitches from the amazing smell of cinnamon and chocolate that had once been baked into the treats I'll probably never get to have again in my life. I wait for footsteps to come pounding towards me but as my mouth waters for cookies, there is only silence. I can't take the smell for much longer so when my stomach rumbles louder than my pounding heart I scoot out and crawl from one empty table to another until I reach the produce stands and the smell of rot wipes away memories of cakes and pies.

Silence, no one's here. I pull myself to my feet and make a face at the nasty smashed and rotten goo that's stuck to my hands and knees. I scrape it off and head out of produce deeper into the store. Wreckage fills every aisle I look down. Shelves have been swept bare and some have been toppled over. Anything that can't be eaten or used to survive has been trashed on the floor and stomped on. It so different from the last time I walked these aisles that I just gape in amazement. A cold breeze hits me as I round a corner and the light gets a little brighter. I come to a stop and just stare. Thirty feet away are a set of main doors to the parking lot. I can see even from here that all the glass has been smashed out and one of the automated doors that slide open has been knocked out completely and lies on the floor.

My feet start moving towards them as my eyes stay locked on the churning grey sky. I can barely see through them. It looks uglier the closer I get to the doors. I'm so caught up in the sight of a sky like I've never seen before that at first I don't register a sound that's headed my way. I'm not scared by it because I recognise it. It's a sound I've heard hundreds of times before and one that always made me happy. I step out of the aisle I'm in straight into the noise like a total idiot. The skateboard hits me right above the ankles a split second before its rider smashes into me with a full body tackle. We go flying while locked together and hit an empty shopping cart that digs painfully into my back before rolling away. My elbow hits the hard tile of the floor and a scream of pain flares out only to be cut off as the full weight of the rider lands on top of me making all my air whoosh out of my chest. The back of my head bounces off the floor once and then we come to a stop.

I manage to take in some air when the rider shifts but lose it just as fast when his hand comes down on my belly to push himself off of me. A tortured groan barely makes it past my lips as my eyes flutter open to see who hit me. Staring back at me are bright blue startled eyes that look me over before they crinkle with a grin. I think it's a kid but his head and face are wrapped up in dirty scarves so it's hard to tell. Thankfully, when he grabs my arm to pull me up to a seated position it's not the one that I smashed cuz it's totally throbbing. Once I'm sitting up I shove the ski mask up over my head so I can see the kid better. Now that he can see my face the grin in his eyes widens and he swipes his own scarves off. He's about my age with blond dreadlocks that reach his shoulders.

"Dude! Epic collision!" Comes out of his still grinning mouth as he offers a hand to help me to my feet. I let him pull me up but my eyes track around the store looking for anyone who might be with him. When I'm satisfied the kid is alone I help him dig through a pile of back to school binders that we knocked over in the collision to find his skateboard. He checks the wheels to make sure they're still working before holding the board over his head in triumph.

"I live to skate another day!"

I can't help but crack up at his drama, it's such a relief to meet another kid and not a crazed looter. He shoots me another cocky smirk and holds his fist out for a bump.

"Sorry about the damage man. I thought this place was empty. I'm Marsh."

I bump him back.

"Rex, no worries I'm padded pretty good. What are you doing here?"

Marsh looks me up and down with an intense stare before his shoulders slump and his eyes go sad before they hit the floor.

"Looking for medicine, my dad's sick. I tried the drug store too but it's totally bare." He looks up and shrugs, "You?"

I flash to the bags of pill bottles and blister packs Daniel had given to Belle from the pharmacy but play it cool. I don't know this kid and he seems cool but I've got the girls and Matty to take care of so I just say, "Water."

He nods in understanding and we just look at each other for a minute before I ask what I really want to know.

"What's happening out there? Is it safe to move around? Is there anyone in charge helping people?" I try to be cool but my voice cracks on the word helping.

Marsh's face is confused and he starts shaking his head.

"Dude, where you been? It's totally messed up! There isn't anyone helping anyone anymore."

I look away from him and make up a story on the spot. "Yeah, that's what I figured, just wondered if you'd seen anything different. I've been hiding out since it all started and this is the first place I came to, to look for supplies."

Marsh pats me on the arm in sympathy. "Sorry to tell you Bro but things are seriously effed up. Pops made us stay in the basement for the first few weeks but we came out two days ago to check things out. There's bodies all over the place and they look pretty nasty. We found a group of people living in one of the hotels but the leader's a major league spaz. He told us if we want in we've got to give him supplies for a room. Dads and Pops are considering it cuz they've got water and a generator." His face and tone changed to anger. "This sucks! I just want to go home! We don't even live here. We rented a house for a few weeks so Pops could go bow hunting. I think we should just walk home to Cali but now Dad's sick, nothing works and we're stuck here! Total life fail!"

I felt bad for the kid and knew how he felt being so far from home. We walked around the half empty store kicking junk out of our way as Marsh told me his story.

I thought he was with his dad and grandpa but turns out he has two dads that are married. He calls one Dads and the other Pops. His Dads is a doctor and Pops is a retired soldier that likes to hunt. They live near San Diego and came up here so his Pops could scout for a big bow hunt that happens in November. They were just having a family holiday when all this madness started. He told me more about the hotel group and how they had sneered at his two dads. Marsh said that they were used to that kind of prejudice but now they were afraid to trust anyone. His Pops wants to find more people to group together for safety but they didn't know anyone here in the town and now that his Dad was sick they couldn't move anywhere else.

I dodge any questions he asks me and think hard. I think his Pops is right about safety in numbers. I think about us three kids with just Belle to protect us and wonder what would happen to us if she got hurt. Marsh seems like a cool kid and if his dads are a doctor and soldier then they would be good people to team up with. We can't stay in the staff room forever and if anyone comes looking back there for supplies we might get hurt. I should go talk to Belle about this but then I'd give away our location so I take a deep breath and ask, "What kind of medicine does your dad need?"

He says a name that's Greek to me but I repeat it over in my head so I won't forget it.

"This medicine would have been in the pharmacy here?" At his nod, I look away and consider it again. If things are as bad as he says they are out there then this might be our one shot to team up with decent people. None of us are from here so it makes sense to group up. I nod to myself and take a deep breath.

"I think I can help you get what you need. Can you go get your Pops and bring him back here? I'll go see what I can find and meet you guys back here."

Marsh's face is filled with hope and excitement for a split second before it turns to suspicion and he asks,

"Why? Why would you help us? What'd you want? Nobody does anything for anyone anymore for free."

I look away and scuff at the floor with the toe of my shoe. He's right, I do want something. I want help. I don't want us to be alone anymore. Should I tell him the whole story or wait till he gets his Pops to come? I want him to trust me so I give him a little of it.

"We don't live here either. My Mom, baby brother and I were in a car accident on the highway that day. She…she died. A lady and her daughter helped us get back to town. We've been hiding out ever since. We managed to get some supplies and medicine but we're alone too. I think we should help each other."

I say nothing about Daniel Ross or the things he told us about what happened or where we are hiding. That's something I need to talk to Belle about first.

Marsh's face changes to sympathy and he gives my arm a soft punch.

"That's harsh bro. I'm sorry about your mom." He chews on the end of one of his dreadlocks and then nods. "Strangers in a strange land, yeah, we should hook up. Go talk to your lady and I'll go get Pops. Meet back here in two hours?"

At my nod, he offers another fist for a bump and then drops his board and skates towards the front of the store. I just stand and wait for a good ten minutes to make sure he's gone before I head back to the staff room. I'm nervous about this and what Belle will say but we have to make some moves on our terms before it's too late.

Chapter Eleven-Skylar

"Square your hips, relax your shoulders and remember to breathe. Don't jerk the trigger. Squeeze it gently back and remember to keep your eyes open!"

I huffed an impatient breath out. It had been almost three months since Dad had started to teach me and he's finally going to let me shoot. My aim with the crossbow has gotten much better but I still struggle with re-cocking it. I would love to use the long bow but my skinny toothpick arms can't handle the draw. Dad made me a smaller version of it but every time I tried to shoot it, I felt like an idiotic cupid!

I can now strip down, clean and oil almost every type of weapon we have. He won't let me touch some of his favorites but he says he will in a few years. I've dry fired most of them and practiced my stance over and over. I'm ready but he's driving me crazy with his nerves!

"Dad, I got this. Let's go already!"

"Ok, ok, we'll start with the .22 Long Rifle semi-auto pistol. It's lightweight and has an easy recoil. Just remember everything you've learned. I moved the targets up to ten feet so you can see how your shot placement is but we'll move them back as you get better."

I gave a firm nod of determination and went over the checklist in my head one more time before taking a deep breath and slowly let it out. The first shot was a bit of a jolt but I knew it was coming so it wasn't that bad except for the part where I missed the target. My second shot missed again but the third one winged the paper and the next few were hits also. By the time I had reloaded twice more I was consistently hitting the paper and more and more were inside the silhouette. When I finally put the gun down my shoulders and hands ached but I felt amazing. I turned to Dad with a grin but it slipped away at the sadness on his face. He pulled me into a hug and kissed the top of my head.

"You did good, Sky. I wanted something different for you, something peaceful, but you'll make a great soldier one day."

It took a long time but every day I got stronger and my shooting improved. I liked shooting the long rifles the best and hated the bow the most. I kept at it but I never really got good at it.

Days went by and then months and then it had been a year since we came here. It's a sad and happy day. Mom and the world died a year ago but it's also Benny's first birthday and I know Mom would want it to be special so Dad and I try hard to be happy for him. I smile and sing the birthday song but inside I'm thinking about all the things this boy will never know.

After the party has been cleaned up and Benny is sleeping for the night, Dad and I just sit and stare at the TV, not really watching the documentary about ancient Rome that's playing. AIRIA has almost every TV show in her data banks but it's too hard watching stuff about the world we used to live in so we stick to history or educational shows. I'm jolted out of my memories of Mom when he speaks.

"Tomorrow when Ben is napping, we're going out."

My jaw drops open in disbelief.

"What do you mean, out? Like out, out? Like outside?"

He just looks at me and nods.

"How, how is that possible? You said it's not safe! You said that there's radiation and that we'd have to stay in here for years!"

"Calm down, honey. AIRIA has sensors all over the mountain and also monitors the few satellites still in orbit. Yes, there is radiation but it's not concentrated around here. There was fallout and there are waves of radiation that travel around the earth. The sky and air are still filled with contaminants from the ash and smoke that was kicked up into the atmosphere. We will be wearing masks and containment suits. We'll be fine!"

I stare at him in shock and terror for a second before I jump to my feet.

"No, NO we can't go out there! What if something happens? What if we get hurt or die, then what happens to Ben? No Dad!" Dad gets to his feet and puts his hands on my shoulders.

"Honey, nothing's going to happen! We will be in constant contact with AIRIA the whole time and she will be monitoring Benny. She'll let us know if he wakes up while we're out there and we'd come right back. Trust me, I've already been out a few times and nothing's happened. You need to start going out so you can get used to it. One day we'll have to live out there and start rebuilding."

My head is shaking back and forth in denial. He's gone out? He's been out there? He left us here all alone? I'm so scared and mad and I feel betrayed somehow that I wrench away from his grip.

"No Dad! YOU go! I'm staying here with Benny. I won't go out there and leave him alone!" I spin away from him and rush into my room where I throw myself on the bed and shake and cry. I can't go outside, ever!

The next day I stand listlessly as I watch and listen to him explain how to suit up. He goes over the equipment and how the decontamination works on the way back in but it's through a fog. I feel like he's abandoning me. He holds out what looks like a cell phone but a little bit thicker and explains that it's AIRIA. She'll be monitoring him the whole time and that I can contact him through it if I need him. I just nod and stay staring at the door we came through so long ago, even after he's kissed me goodbye and it's slid closed. Eventually, I sit on the couch but I keep my eyes on the door and wait for him to come back.

It gets easier, every time he goes out and comes back but it's still hard for me. I try and not let him know how much it worries me because I see how much he needs to go out sometimes. He's happier after he's walked the mountain. This place must feel like a prison to him but it stopped being that to me, to me its home and safety. He shows me pictures of a dark grey sky with boiling clouds and dying trees and I don't know how being out there can give him peace because all I see is what we've lost. Twice while he's out there the alarms go off and AIRIA announces that a radiation cloud front is moving in. I have to sit and listen to her count down the minutes until it reaches us. I have to listen as she counts down the temperature drop as the world outside freezes solid, but both times he makes it back safely.

It gets easier, until one day it gets really hard.

"Daniel Ross, no life signs detected. Skylar Ross, access level increased to green."

My head snaps up from the textbook I've been reading and my eyes shoot to the door. What, what did she say? It's like an echo of what she said when Mom died but that can't be right! I look to Ben who's playing with wooden cars Dad carved for him then back to the door before whispering.

"AIRIA, what did you say?"

"Daniel Ross, no life signs detected. Skylar Ross, access level increased to green."

I shake my head no and whisper again.

"Where is he?" When the computer doesn't respond I try again. "AIRIA, where is Daniel Ross?"

"Daniel Ross, last known location is zero point eight kilometres north from main access door."

My head keeps shaking in denial and my eyes stay glued to the door. That's not far from here. He'll be coming back any minute. I'll just watch the door and it'll slide open and he'll come back. He always comes back. But the door stays shut and every minute that goes by is like a fist pounding my chest until a tug on my sleeve finally pulls my eyes from the door.

"Ky, was green? You no green!"

I look down at my two and a half-year-old brother through a blur of tears. Green, green means we're alone. Green means I'm in charge. Green means we're orphans now.

I just turned thirteen two weeks ago and green means I’m all he has left.

Chapter Twelve-Rex

"Are sure about this Rex?"

Belle and I stand in the wreckage of back to school supplies and wait for Marsh and his Pops to arrive. We've gone over and over the situation and what would be best for all of us but we're both still nervous. If Marsh's Pops isn't a good guy then we're hooped. We plotted out an escape plan but it mainly comes down to split up and run and if this guy is a hunter like Marsh said then I doubt we'll get very far. We've just got to hope we can trust them.

I shrug my shoulders in response to her question but we both crouch down when we hear footsteps. We watch the main aisle coming into the store over the top of a shelf of pencil cases. First, a man comes into view and even in the dim light we can see he has a large compound bow that he points down every cross aisle he passes. Then Marsh comes about ten steps behind him. I glance at Belle but she's locked onto the guy and as he gets closer she calls out.

"That's far enough!"

The guy freezes in place for a split second before pointing the arrow he has notched in our direction. He holds up a fist and Marsh comes to a stop too. No one says anything so I look at Belle and nod my head towards the guy but she's just frozen. I'm about to call out but the guy beats me to it.

"My son Marshall has said you might be willing to trade for some medicine. What do you want in return?"

Belle's still looking like a deer in headlights so I slowly stand up with my hands above my head.

"Can we talk? Without the big pointy stick aimed at us?"

I hear Marsh snort a laugh and take it as a good sign so I pull Belle up from her crouch and we step out from behind the shelf. The guy looks us up and down then scans the area around us before lowering his bow and waving us towards him. At the same time, Marsh moves up and stands next to his Pops who motions him back with his head but just gets an eye roll from his son who stays put.

"Hey bro, this is my Pops. I told him about what we talked about. Is this your lady?"

I smirk at Belle's disbelieving look and end up laughing. It breaks the tension for all of us so Belle holds out her hand to the guy and introduces herself.

"Hello, I'm Belle and this is Rex. Your son has told him about your situation and we've talked it over and think we should work together. We were also stranded in the area and don't know anyone locally. There's something to be said for safety in numbers."

The guy shot a frown at Marsh before nodding his head.

"I'm Lance. Marshall told me you took charge of Rex and his little brother. Is that true?"

Belle shrugged. "Their mother had died in a crash on the highway. I couldn't just leave them out there alone."

"So you made it to town here and have been in hiding ever since? None of you are sick?" When Belle shakes her head he asks, "Why did you hide? Why didn't you look for other people right away, like the authorities?"

Belle looks at me and bites her lip so I nod at her to go ahead. We have to trust someone. So Belle tells them about Daniel and everything he told us and all the advice he gave us. The only thing she doesn't tell them is where we're staying or how many supplies we have. We have to hold that back until we have a deal.

"All of that sounds plausible. He gave you good advice, you're lucky. A lot of people didn't take shelter and they died. We stayed in a basement until a few days ago, but my husband's sick and needs antibiotics. There's a group at one of the hotels but they're not real friendly and they claim they have no medicine. I've looked through the pharmacies but they've been cleaned out so I've been trying empty houses with no luck. Apparently so has Marshall." He shoots a stern look at his son who just shrugs.

Belle pulls a pill bottle from her pocket and hands it out to Lance. "Here, this is the medicine Marsh said you needed for your husband."

Lance's gaze goes from the pills to Belle's face. "What do you want for them?"

She shakes her head. "Nothing, take them. I would like to talk some more though about our situation and how we can help each other. We don't know anyone and we don't know who we can trust. I'm alone with three kids and one of them is just a toddler, I don't know if I can do it all by myself. If you'd be willing to team up there are things we can share."

Lance stares at Belle hard in consideration before his face softens and he reaches out to take the pills.

"Ok, we don't know who we can trust either but we have to start somewhere. So, do you have a secure place to stay? If you don't, you and the kids should move in with us. It'll make it easier to protect you. It will also help to find food faster if we work together."

Belle looks at me and raises her eyebrows in question. I glance at Marsh who shoots me a wink and thumbs up before finally looking at Lance. This is a big risk but I just don't see us making it on our own so I nod and say,

"Actually, we think you guys should move in with us. There's no way we could move all the supplies we have without anyone seeing them and coming after us for them."

At Lance's confused look we tell him about the pallets of food that's still stacked up in the back room. His eyes get real big and then panicky as he looks around the store.

"Holy crap, you guys are on borrowed time! People will be coming here to search for food. It's been just over two weeks and most of the people left alive are starting to move around again. They would have gone through most or all of the supplies they looted on that first day. The first wave of death is over, now it's really going to get ugly! You don't want to be anywhere near this store when that happens!

Marshall, take these pills to your dad and tell him to start packing, that we're moving to the fallback point tonight, then get back here as fast as you can to help.

Rex, please go to the outdoor and gardening section and see if there's any utility wagons left. There might be some that haven't been assembled in boxes. Oh, and see if there's any seeds left! If there is, bag as many as you can carry and bring them to the back room.

Belle, can you show me around the back of the store? I need to get an idea of just how much we can take with us."

We all just stand there nodding until Lance barks, "GO!"

And then we scatter.

After that things move very fast and frustratingly slow at the same time. Lance has us all moving at packing up the staff room and stacking things by the exterior door we had first entered the store through. He's moved the toaster pallet as well as others in front of the main store doors closest to the staff room to create a better barricade, but now we're all terrified that looters will be pounding them down at any minute or attacking through the doors we left open at the other end.

As the girls pack up the staff room, he has Marsh, once he's back, and I take overlooked supplies from the main store. It's things Lance calls comfort items like more clothes, bedding and beauty supplies. Basically, anything people would be willing to trade for after food and water. Lance is assembling the three garden trailers I found still in their boxes to help haul as much as possible. They're not very big but those three plus the bigger display model that was in the garden center will help us move more than we could ever carry.

The plan is to get everything we want into the back room and then barricade the other set of doors so we have some protection if anyone comes to search the store.

When we finally haul the last trailer down to the staff room it's late afternoon and I'm wiped out. I've done more today than I have in the last sixteen days and all I want to do is collapse onto my air mattress. The massive amount of supplies that now fill our end of the store room is staggering. There's no way we can move that stuff with these four puny wagons, but Lance is determined and he calls us all around.

“Alright guys, I know you’re tired but we have to get this done. I’m going to leave now and scout out a route to our new home while there’s still some light left. After that, I’m going back to our place to grab Ethan, Marshall’s Dad, and move our stuff over to the new place. Once I get back, we’ll wait for dark and start moving this stuff. We’ll have to haul all night. From what I’ve seen so far, no one comes out at night. With the street lights out and the heavy cloud cover blocking the moon and stars, we should be able to move a lot of these supplies without being seen.

I need you guys to load up the trailers so they’re ready to go as soon as possible. It’s going to take us many trips to get this done. We will only be able to take three on the first trip. Belle and I will pull one each, Rex and Marshall will try and take the third and Sasha will push the baby in the stroller. Once we get to the house, we’ll leave Matty with Ethan and we’ll come back and do it again, this time with all four trailers. Try and pack the lighter stuff on the trailers you boys will be pulling. Sasha, you’ll be helping push for the boys when they need it.

This is going to be hard but the more we move, the better chance we’ll have to survive.”

Everyone nods their agreement so Lance slips out the back door and Belle re-locks it. She turns to us and goes to speak but after looking at all our faces in the dim lantern light shakes her head.

“Ok, first things first, dinner! It will take Lance a while before he gets back so let’s have some dinner and then we’ll get to work.”

I let out a groan of relief and Marsh jabs a fist in the air.

“Oh man, a meal not made by the Dads, righteous!”

I snort out a laugh. As much as I’m dreading the night to come, I’m happy we aren’t alone anymore.

Chapter Thirteen-Skylar

It's been three days. Three days of Benny asking for Dad and me having no answer. I've barely slept, barely eaten. I've been like a zombie doing the bare minimum. Milk the cow, collect the eggs, feed the baby. I don't know what else to do. I need to know for sure but that means going out there and checking. Maybe he lost his communicator and that's why AIRIA can't get a reading but then why hasn't he come back? I need to know but if I go out there and find him then that's it, it's over, final. Right now there's still a chance Benny and I aren't alone but if I check, that chance will be gone and I'll have to face the reality that we're on our own.

I angrily brush away the tears that keep leaking down my face. I have to know.

Benny's breathing levels out and he's finally asleep for his daily nap, so I rush to the door and palm it open. Inside is a small closet that has the containment suits that I hadn't noticed the first day we came here. Dad had planned on taking me out so he had customized the arms and legs of the suit to fit me but there's no way to change the dimensions on the helmet. It's too big for my head and it won't stay on properly.

More tears leak down my face but this time they're in frustration. I have to know!

"AIRIA, can I go outside without a suit helmet?"

"Skylar Ross, radiation levels are at acceptable levels for nonfatal doses. Ash and air contaminants are present. A filtered face mask is advised."

Right, face mask, ok. I shove hanging suits aside until I find what I think will work and then tighten the straps to keep it on my face. I grab a communicator and turn it on.

"AIRIA, I'm going outside. Will you give me directions to my dad's…" I choke on the word body and have to take a minute to get it together before I try again. "AIRIA, Please give me directions to Daniel Ross's last known location."

"Skylar Ross, exit decontamination room and proceed north for eight hundred meters."

North, right, except which way is north? I growl in frustration.

"AIRIA, is north right, left or straight ahead?"

"Skylar Ross, north is to the left of the entrance."

Left it is! I take a deep breath and ask her to open the exit door then just stand there staring out. The sky is so ugly, every shade of grey to black rolls through the heavy clouds. When I finally tear my eyes from it and look around the clearing it's not much better. The trees, the grass, the rocks are all covered with what looks like dirty snow but I know it's ash. Ash that is all that remains of burnt cities and people.

I don't want to go out there. My breath is heaving in my chest and my blood is roaring in my ears but I have to know so I step out. The door whooshes closed behind me causing me to flinch and making me want to claw it back open.

I have to know!

I turn left and start walking. Every step crunches the brittle dead grass under my feet. My eyes are constantly moving looking for a threat of some kind but I see nothing except a dead grey world. There's a bleak winter wind that makes my skin shiver under the suit. I didn't expect the cold. I didn't expect to feel this hollow.

My feet keep moving as I go around trees, bushes and rocks but always north. I'm on an incline that makes my legs burn. It's been two and a half years since I've had to walk on anything but a flat surface and over a year since I've climbed stairs to the lower level. My legs tremble at every step and I keep my eyes down on the ground in concentration until I reach level ground again.

And then I stop. And then everything stops. I no longer feel the burn in my legs or the cold biting wind because there are his feet and they are blue. My eyes don't want to move from those bare feet. My mind doesn't understand why his feet are bare.

I have to know so I look. I look at all of him. He's not wearing anything except his boxer shorts and he's blue and in the middle of his back is one small hole.

My knees give out and I hit the dirt beside him and go away for a while.

A horrible noise brings me back and it takes me a few seconds to understand that it's coming from me. I sound like a wounded animal. My hands reach out to try and turn him over so I can see his face but he won't move. He's frozen to the ground. His arm feels like solid rock through my gloves and that ugly sound becomes louder.

I can't leave him like this out here in the open but I can't move him. I tear my eyes away from him and look around for his clothes. Why isn't he wearing his clothes? Everything is gone. His suit, his boots and socks, his rifle and mask, even his communicator are nowhere to be seen. I don't understand what this is, how this happened until my eyes go back to him and I focus on the small hole in his back. Then everything clicks into place.

MURDERED! My father was murdered. Someone shot him in the back and then stripped him and stole all his clothes and gear. My first reaction is fear. I jump to my feet and spin around. What if they're still here? What if they shoot me too? I'm such an idiot! All that time learning how to use a gun and I didn't think for one second to bring one with me for protection. I'm seconds from launching myself back the way I came and to the safety of the cave when I remember it's been three days and whoever did this is long gone.

Next comes anger. I scream in rage. Why, why did this happen? Why has everything been taken from me? It's not fair! First my home then Mom and now Dad!

I kick out and my boot slams into a rock. The pain in my foot settles me down but I have to rip off the face mask to wipe away the tears and snot before I can breathe properly again. I get my mask back on and look down at the rock I kicked and then back to him. I can't leave him like this and I can't move him but I can cover him.

Rock by rock, I slowly cover him. It takes a long time to hunt down and carry back rocks that I can lift but I never stop. I think about all he did for us and how he's with Mom now. I think about who did this to him and if I'll ever know. Rock by rock I get colder and harder inside until AIRIA speaks.

"Skylar Ross, Benjamin Ross has awoken and is calling for you."

"Thank you AIRIA. Please tell Ben I will be there soon. I have found my father's body but all of his clothing and his rifle are gone. Even his communicator is gone."

"Skylar Ross, current GPS location of Daniel Ross's communicator is thirteen kilometers south of your present location."

"What? What does that mean? What is thirteen kilometers south of here?"

"Skylar Ross, the town site of Canmore is at that location."

Like a knife cutting through me, I remember yelling at him, begging him to help those people. I remember him telling me we couldn't, that it was too dangerous, that we had to just keep our family safe. I remember how mad I was at him, how disappointed I was in him.

As I place the last rock on his grave, the rage comes back but this time is not red hot. This time it's cold white. All those people I wanted to help. I hope they all burn and die because now I know.

Chapter Fourteen-Rex

That was full-on terror. Thirty-three minutes of pulse-pounding terror and now we have to do it again and again and however many more times it takes to move the supplies. It's so dark. Not like dark where you can stumble around and find your way dark, dark like can't see my hand in front of my face dark. The only thing that guided us was two small flashlights that still work, one at the front of the line and one at the back.

The trailers weren't that hard to pull once we got them moving but my shoulders and back say different now that we've stopped. We're in a house behind a house. Lance's friend who invited him up north owns this huge monster vacation home in town. It's where he would have stayed when he came up in November for the bow hunt. We aren't staying in the monster house but the original home that was built on the property. His friend told him the story of how the property originally belonged to his grandmother who passed it down to him when she died. He wanted something new and modern but the original house on the lot was where his mother grew up so he restored it as a guest house for her and built the new one in front of it. The reason Lance picked it out for us to stay in is because his friend kept all the old fashioned features like the three wood burning fireplaces, a hand pump in the kitchen as well as the wood-fired cook stove, alongside the modern plumbing and heating. He said it was charming. For us, it means well water and heat. It also has the bonus of being set back from the road and hidden by the main house which is surrounded by a fence with a gate.

My hands and feet are numb with cold by the time we unload the last trailer and all I want to do is crawl under a blanket and sleep, but there's no chance of that happening with all the supplies that still have to be moved.

Marsh's Dad, Ethan seems like a good guy. He doesn't seem to be as hard as Lance but that might be because he's sick. The guy has a pretty bad infection of some kind but he still kept trying to help us unload until Lance put him in charge of Matty. He must be good with kids as Matty fell asleep in his arms pretty quick, or it just might be that its hours after his bedtime and the heat from Ethan's fever made him sleepy.

I'm just grateful he's here to watch my brother while the rest of us make the trips back and forth from the store. They both sleep through our coming and going four times but Ethan comes out of the bedroom on the fifth trip and demands that we stop. It might be because there's a hint of light in the sky or maybe because the three of us kids are staggering around banging into things in exhaustion. Whatever the reason he's my new favorite person. I can't even see straight as he leads Marsh and me into a room and pushes us onto a bed. I'm out even before I feel the heavy blankets settle down on top of us.

The next time I open my eyes I think I must have only slept for a few minutes cuz my whole body aches and the room is very dim, like at the start of a day. It's weird waking up to natural light after being in the staff room for so long with candles and the lanterns. All I want to do is roll over and close my weary eyes but when I do I see that Marsh is gone and then I hear Matty laugh somewhere in the house. As much as I want to sleep and am thankful there are other people to help with him, Matty's my brother and my responsibility. I push myself out of bed and stand bent over to stretch my aching back and wonder if this is what it feels like to be really old. I shuffle towards the door and almost face plant when I trip over my boots. Huh, someone must have taken those off me last night. When I finally reach the door and pull it open, the most heavenly smell hits me right in the face, making my steps quicker. I didn't really get a good look at the house last night so I glance down the hall and see two more open doors with bedrooms and another one with a bathroom behind it. Turning the other way leads me out to the main living area where the living room, dining room and kitchen all flow together. Everyone's sitting around the dining room table eating and as I get closer my eyes are drawn to what my nose told me I would find. Bread! Two loaves of what can only be freshly baked, unsliced bread sit in the middle of the table.

It's so easy to take things for granted when you have them every day but take them away and it's kind of like grieving except you don't know how bad you feel until it's there again. I feel sort of dumb but my eyes get a little damp at the sight of that bread.

"Dude, drool much?"

I look up from the bread to the laughing face of Marsh and reach out my hands making them into lobster claws.

"Gimme!"

Everyone laughs and Belle cuts off a few slices for me. She passes me a plate with a smile.

"We thought you might sleep right through the night as well as the day!"

Wow, it must be almost night again. I guess I slept more than I thought but I just shrug cause my mouth is full of heaven. Lance stands up and waves me into his chair so I drop into it gratefully. I just can't believe how sore my body is.

"Alright, now that everyone is present, we should go over our next steps. We made a great deal of progress last night but we didn't get everything so I'd like to do a few more runs tonight if no one has gone in there today. I'll leave to scout it out in a few minutes and if it looks ok I'll come back and we'll head over after dark.

I took another look around the neighbourhood earlier today and as far as I can tell, there's only two other homes on this block that are occupied. This block seems to have a lot of vacation homes so we're lucky that most of them were empty when everything went down. We still need to be very careful. I'd like us to remain hidden as long as possible so we'll have to board up some of the windows and cover others with heavy curtains to keep the light from showing. We will need to set up a watch rotation and work on securing the gate and fence as best we can.

There's a lot of work that we'll need to get done in the days to come. Half the supplies should be moved to different locations so if we're compromised, we won't lose everything. We need to start working on growing some food. The low level of sunlight will make that a huge challenge but there are some things that can grow in these conditions. We can also help that along with setting up a few grow lights"

Belle interrupted Lance with a frown of confusion.

"Wait, grow lights? Does that mean the electricity will be coming back on soon?"

Lance shook his head sadly.

"Sorry but I don't think that will be happening anytime soon, like years. BUT, we can make some of our own. It wouldn't be enough to run an entire house full of electronics but it would be enough to power up some grow lights for a few hours each day. We can set up a line of small windmills that run to a bank of batteries but that's a project to think about down the road. We'll have a lot of scavenging to do after we get this place set up and secured.

Ideally, we want to stay off the radar for as long as possible. People will be more desperate and dangerous during the next few months. After that, the only ones left will be groups that have set up sustainable shelters. Then we will make contact with a few, if they're friendly, and talk about possibly trading."

Lance took a good long look at all of us. It was a lot to take in but I was feeling pretty relieved that we had hooked up with him and his family. He seemed to know what to do and the plans for the future were comforting. I looked over at Matty who was decimating a piece of bread and breathed a sigh of relief. The kid had a way better chance of making it now. I turned back to Lance when he started talking again.

"Take a good look around at the people here. This is now your family. We will be sad, mad, frustrated and irritate the hell out of each other but none of that will matter in the end. All that will matter is WE ARE A FAMILY and we will stick together, have each other's backs always, and most importantly, SURVIVE, for as long as it takes."

Part Two

Chapter Fifteen-Skylar

"Skyyyyyy-larrrrrr!"

I grit my teeth and hit the punching bag harder, hoping the next words won't follow but knowing they will and sure enough…

"I'm bored!"

There they are. Rather than take my frustration out on my seven-year-old brother, I execute a powerful roundhouse kick high up on the bag to finish off my session in the gym. I walk around the gym in laps until my heart slows down from the workout and mentally count to one hundred. I only manage to get to seventy-four before it comes again. This time with slightly more whine to it.

"Skyyyyyyy-larrrrrr!"

I take a gulp from my water bottle and snag a towel to wipe my face, stalling for a few more minutes of not being Ben's mom slash big sister. The boy amazes me every day. Most days I love him so much it hurts and I am just so proud of the incredible little person he's grown into. Other times, I want to throttle him into silence. I get now why Dad went outside those first few years and always came back happier and more peaceful. I started making my own short trips out when Ben turned five. AIRIA turned out to be a pretty good babysitter and bribery goes a long way in getting a five-year-old to do as you say. We made a deal that he would sit on the couch for an entire Disney movie while I went out and in return, he gets half a bar of a rapidly dwindling supply of chocolate bars. AIRIA monitors his every move and if he gets up for any reason she reports it to me through my communicator. If it's for any reason other than a bathroom break, she patches me through to her speakers so I can talk to him. There's never been a problem because Ben's a really good kid and he's desperate for chocolate.

I never go very far. The second time I went out I found a hunting stand up in a huge tree that I assume Dad built. It's only a few hundred feet from the main entrance door but it might as well be miles away. I swear sitting up in that tree every other month for roughly ninety minutes at a time is the only reason I haven't gone completely fracking insane. I'm always armed with one of Dad's favorite rifles but I never shoot it, even when the few animals left alive wander under me. I carry it to protect myself from a different sort of animal but thankfully, I've never seen one of those. As an added safety precaution, AIRIA now monitors the surrounding area for movement of anything with a heat source that weighs more than a hundred pounds and walks on two legs. So far she hasn't detected anyone anywhere near our doors.

Just being out under the sky with no noise but the wind in the dead forest helps me find a little bit of peace so I can go on being all that Ben needs me to be. The only other time I feel anything close to that peace is when I run or practice kickboxing. I don't have role models or parents or anyone to help me deal with all that's happened or all the ways I'm changing as I grow but I did find some help in a weird way.

After Dad died, it was really hard for me to deal with a two and a half-year-old on my own. He needed so much time from me for everything, including entertainment that I was losing it bit by bit. One day I got impatient and told him to go watch TV. At the poor kid's confused look I realised that he had never watched a TV show before. He was always asleep when Dad and I would watch a show at night. I asked AIRIA if we had any kid's shows and hello big eared mouse and his pals! I had never seen Benny sit still for so long before, he was totally enraptured. Two hours of Benny in a TV coma a day made a huge difference in my patience levels. I could actually get my chores and a few other things done.

The thing about TV is it can be totally addictive. With Dad not around to pick out what we watched I was free to choose my own shows. I don't even know why Dad bought DVDs because AIRIA has like every season of every show in her data banks. It's like Netflix on steroids. I started out watching all the Family channel shows I used to watch but they just seemed so young and silly to me that I wasn't even laughing at any of the jokes or pranks. So I moved on to teen dramas. I learned what fashionable clothes I should wear if I lived in Manhattan and how to stab my BFFs in the back at every turn. Two brothers taught me how to kill every supernatural creature ever thought of and how to end many, many, many apocalypses that could happen. I learned what the earth will look like after a nuclear war if I live to be a hundred and that there would be a war between survivors on the ground and kids from space. Ironically, I don't know who will win because you know, nuclear war killed that show.

Watching all those people do their thing both filled a void and made it bigger. I'm never going to wear a prom dress or break up and makeup with a best friend or a boy for that matter, so I lived through those characters and it hurt a lot. The show that had the most effect on me was about a group of people who were forced to live in space ships after robots that looked like people, destroyed all their planets. BEST SHOW EVER! This girl is so sad and so fierce all at the same time while mourning all that she's lost and searching for a home to have a future on. I related to her so much and wanted to be strong like her. If Ben hadn't been here with me, I probably would have opened the alcohol stores and drowned my pain away like she does. That character pushes through her pain and she uses exercise to help do it. So I try and be strong like her even though I don't have a really cool call sign.

I finally went back into the studio that Dad had set up for me and put it to good use. I watched videos that taught me how to fight. Kickboxing was my favorite and then I started running on the treadmill which was just about the most boring thing in the world. My growing body ached to move, to run but not on a machine that went nowhere. I needed space to run but doing laps around the cow pen was out and then I remembered.

After Dad died, AIRIA had changed my access level to green. It took a few weeks after I found his body for me to remember that. I was stumbling around in a grief-filled fog when I walked passed a door that I had never been able to open because of my access level. I'm not sure why I had even tried the door again after all that time but it slid open when I palmed the plate. I stood on the threshold and tried to see in but it was pitch black.

"AIRIA, why did this door open?

"Skylar Ross, access level green."

The thought of all the doors that wouldn't open for me before being opened made me pause. How much more was there to this place?"

"AIRIA, are there lights in this room?"

"Skylar Ross, lights are motion activated. Please proceed for illumination."

I took a nervous step into the darkness and blinked hard when fluorescent lights lit up on the ceiling. My mouth dropped open in astonishment when lights kept coming on down a very long concrete corridor. From where I was standing I couldn't see any doors or breaks in the walls that ended in the distance at a set of double doors. My feet moved hesitantly down the corridor toward the far doors. I was sort of scared but a flicker of excitement was growing inside of me, temporarily pushing aside my grief.

Standing in front of those double metal doors, I wondered if what I was about to find would change everything for me.

They slid open just as easily as the first door and my movement through them brought more lights. Slightly disappointed but still curious, I poked my head into plain offices with boring desks and computers as I made my way along the next hallway. Another set of double doors barred my way but this time I didn't hesitate to palm them open.

It felt like it took forever for the lights to stop coming on this time as bank after bank lit up. I just stood there shaking my head. It was another cavern. Only this one was almost as long as a football field. Row after row of double-decker bunks filled the huge floor. Each one had a rolled up mattress and clear plastic bag with bedding on it. I walked the perimeter of the cavern wide-eyed. This place made my little living area feel like a tiny box.

The first door off of the cavern I came to had a sign marked storage and supply. I palmed it open and only took a few steps in to activate the light. Tall and empty shelves sat in rows but further back I could see metal transport containers. I didn't bother to check how many or how far back they went. Anger was starting to simmer in my belly as I moved on. The next area had metal railings set up like the ones in an amusement park for people to stand in lines for rides. Built into the cavern wall were metal roll down windows and counters. I counted five before I found an access door marked kitchen. Again, a quick look showed me a huge food preparation area with gleaming stainless steel surfaces everywhere.

By the time I had inspected ten washrooms and shower rooms, a massive grow area and a medical clinic with an attached lab I was furious. My Dad must have known this was here. I had begged him to help all those people and he claimed we couldn't. He was a liar and a coward! Those boys we had passed on the highway, all those people hurt or stranded on the highway and in the town could have been safe here and he had done nothing, claimed we couldn't do anything!

I had walked the entire perimeter of the cavern and ended back where I started. Looking out at all the wasted space and opportunity, fresh tears ran down my face.

"AIRIA, what is this place?"

"Skylar Ross, current location is deemed military barracks."

"AIRIA, how many people can live here?"

"Skylar Ross, maximum capacity is one thousand five hundred people."

Fury and frustration had me clenching my fists but then one image of a bullet hole in Dad's back swept it all aside. I turned away and walked back to my home. Turns out, I didn't find anything that would change things for me after all.

So years later I remembered the mega cavern and it became Ben's and my back yard. I used it to run laps and play hide and seek with Ben. We filled that empty lifeless place with as much fun and laughter as I could give him.

"Skyyyyyy-larrrrrr!"

I throw the damp towel in a basket and head out, my brain scrambling to find something to entertain Ben. He has the same schedule that Dad had made for me with school work and chores but he's still only seven years old and he needs more fun things to do than I can think up.

"AIRIA, what did seven-year-old boys do for fun back before the bombs dropped?"

"Skylar, seven-year-old boys got into mischief! Then they played video games."

I rolled my eyes at the ceiling. Four years alone with AIRIA had given me plenty of time to work with her "personality". She's not quite a full A.I. because she doesn't have emotions but she can learn. It took a while and lots of reminding but eventually she stopped being so literal in every sense. She also stopped using my full name every time she spoke to me. The truth is, I needed a friend and all I had was a computer voice and a toddler so I made it work. The humour she's learned to inject into some of her answers makes her both a lot less annoying and sometimes even more annoying.

"Ha ha AIRIA, do we even have any video games in here?"

"Why yes, we do! All models of pre-bombing video gaming consoles can be found in storage container C-12."

"Thanks, AIRIA.

Bennnnn-yyyyyyyy!

GAME ON!"

Chapter Sixteen-Rex

"Well, that's it, we had a good run. Almost seven years but there's no denying it now. The well is dry." Lance scrubbed at his face with his hands before turning to face all of us. "Honestly, that's three years longer than I thought it would last."

We all stand around the kitchen staring at the sink like it'll magically start filling up with water despite Lance's words until Ethan clears his throat.

"Alright, we knew this was going to happen and we've talked about it. We have two weeks of stored water in the barrels to take us through the transition to the hotel. We'll be fine."

Marsh grumbles in annoyance and I shake my head. "I can't believe we have to move to the hotel. Ted and his men are such asshats!"

"Language!" Belle snaps but she also nods in agreement.

Lance holds up his hand to stop the flood of complaints that are about to pour out.

"Yes, they're jerks but it's not like we'll be going in begging for scraps. We negotiated this move a while back. We have things and skills he wants for his group so we won't exactly be under his thumb the way the others at the hotel are. We'll be sharing their space and water but he won't be ruling us like the others. We need to make this work. We need access to that water."

I glance over at Sasha and frown. She's gone even paler than normal. None of us have what you would call a tan with the sun barely showing through the dead sky since the bombs dropped but with her red hair, she's the palest of us all. I grit my teeth in frustration at the idea of putting her anywhere near Ted and his men. Every time we've met up with them to trade they leer at her and make nasty suggestive comments. She's only just turned fifteen and she shouldn't have to deal with that kind of BS. I turn back to Lance with a hard face and tone.

"I don't like the creepy way they look at Belle and Sasha. We need to make it clear to them that they keep their hands to themselves and watch how they speak to them!"

Sasha's face flames up into a red blush and Belle pulls her closer with a frown. I have to look away at the adoring look Sasha sends my way. She started looking at me differently about a year ago and it's uncomfortable. She's my little sister and she always will be so I wish she'd knock it off. Marsh's muffled snort draws my attention and he bats his eyes at me like a lovesick fool which earns him a solid arm punch.

Lance shoots Marsh a "that's enough look" before turning to the girls.

"You two have nothing to worry about. I've already discussed that with Ted. I've made it clear that they're to keep their distance from the both of you and that the first man to harass you in any way will get an arrow through the throat. He's assured me he'll keep his men under control."

Belle nods in relief but I don't buy it. Ted and his men have ruled over the survivors at the hotel for too many years. They treat those people like slaves and they're used to getting what they want. Marsh and I will have to stay close to the girls to make sure they're safe.

"Will there be other kids for me to play with?"

Matty's bouncing on his feet in hopeful excitement making my heart tug for him. Poor kid might have been safe and fed all these years but he's lonely for other children to play with. We all take turns playing with him and Belle and Ethan teach him school work but he needs to run and yell and play with kids his own age. I put him in a headlock and give him a head rub causing him to belt out a laugh.

"We'll see squirt. There should be a few kids over there but you still need to stay tight with us for safety."

He wiggles out of my grasp with a hoot of triumph and bounces over to a box in a corner that's overflowing with toys. A wave of love fills me. I'm so grateful I have Matty. Belle, Sasha, Lance, Ethan and Marsh are my family but Matty's my blood. He's a reminder every day of Mom.

"Alright, everyone, we need to start packing. This move will take a lot longer than when we came here from the store. The hotel is a lot further away but at least this time we can do it during the day. We're stripping this place bare but we're leaving the cached supplies in the hiding places. We don't want to put all our eggs in one basket. All the plants need to go though so we need to start transferring all the box gardens into containers. We've scavenged plenty of them so we should be able to transfer everything.

Even with the large wagon we built a few years ago it's going to take many trips. We need to do this ourselves. We don't want Ted's people to know everything we have so we cover as much as we can with tarps. They'll know we have a lot but they won't know exactly what we have. I'm going to head over there in a bit to let him know we're ready to move and take a good look at the rooms they're giving us. Once we move the first load over, someone will need to stay over there and keep guard. We might have to live with them but we will NEVER trust them."

It took us five very long and tense days to move everything we wanted into the hotel. It was incredibly stressful being on edge constantly as Ted and his men watched our every move. Marsh, Ethan and I rotated guarding our rooms as the others made trip after trip hauling our supplies. Every one of us had to turn away one of Ted's men at the door when they came sniffing around to try and get a look at what we had, or what they thought they might be able to steal from us. We truly were moving into a viper's nest.

Lance had turned down the rooms Ted wanted us in. The guy tried to put us smack dab in between his men's rooms but Lance said no so our living area was in an unoccupied wing with seven connecting rooms. Before we could move in we had to haul beds out of three of the rooms so we could set up a living room, kitchen area, storage room and a growing room. We took the interior connecting doors off all of the rooms except for the ones we would be using for sleeping and used them to barricade all but two of the corridor doors. Being so far away from the main hotel population meant we would have to haul buckets of water, but it was worth it for our privacy and security. Once we had everything moved in we went to work setting up makeshift wood stoves that we had made from clothes dryer drums. Then we set up our small windmills on the roof above our rooms with the wires coming in through a small hole Lance drilled above the window and to the bank of batteries inside.

We were dependant on Ted for water but we would try and be self-sufficient in every other way. The price we paid for daily water rations was Lance and Ethan's skills. Lance would now hunt for both our group and the hotel and Ethan would take shifts doctoring. The rest of us would continue doing what we had been doing for the last seven years. Belle schooled Matty and Sasha and kept our food growing with Marsh's and my help. The two of us did security and scavenged goods from the town when Lance or Ethan was around to watch over the girls and Matty.

We were all bone tired by the time it was all done and we were sitting down to the first real meal in our new home when a knock came on the door. Every one of us turned to stare at the door. Lance gave a weary sigh.

"I guess this'll be the new normal. Seven years with not one visitor. That was nice."

He pushed away from the table and reached for the compound bow that he never went anywhere without before opening the door. I tipped my chair back to see around Belle and promptly dropped it back on all fours before nodding at Marsh and Ethan and shooting to my feet. They followed suit and we formed a line shoulder to shoulder between the door and the table.

"It's Ted and some of his men," I told them in a low voice.

Marsh and I might not be as strong or as hard as Lance but we both had shot up to near six feet in the last three years. Marsh was all long arms and legs but I had a decent width to my shoulders. Living on wild protein and home grown vegetables without the processed junk we used to eat had cut any excess flab from our frames. Chopping wood daily and hauling everything we needed on our backs had added muscle. We might not intimidate them but we would hold our own if we had too.

"Hello, Ted. What can I do for you?" Lance asked while blocking the door.

"Lance, I just wanted to come by and welcome you all to my hotel. Make sure you're all settling in ok, see if you needed anything."

"Thanks, we're settling in just fine."

There was a pause before Ted spoke again, "Surely you'll invite your new neighbours in?"

I could see Lance cock his head to the side before answering giving me a glimpse of Ted's hard expression.

"Well to be honest Ted, we're not quite ready to host such a large housewarming party. We're also just sitting down to dinner."

Again there was a pause as if Ted was waiting for Lance to say more. He finally gave a fake laugh.

"Don't worry about it, we won't keep you long from your table. How about just me and Mickey come in? It's only the right thing to do. As your host, I feel it's only proper for me to welcome your people personally."

I saw Lance tense up but he stepped back and allowed Ted and his number one man into the room before shutting the door on the other smirking men in the hall.

I met Ted's eyes with a level, blank stare. The last thing we wanted was a confrontation so soon after moving in. There was nothing physically special about the man. He was average height with mouse brown hair and plain brown eyes. It was the look in those eyes that put you on edge. They were always looking at everything and they gleamed with calculation. His number one man, Mickey was another story. Everything about the guy screamed "thug". Bald on top, with long greasy hair that hung past his shoulders. His skin was heavily tattooed and exposed as he always seemed to be wearing a tank top. In the frigid cold we lived with every day it was a stupid way to project just how tough he wanted everyone to think he was. He had beady, watery eyes that always seemed to look at you with contempt.

Ted pasted a slimy, insincere smile across his face as he stepped deeper into the room.

"Boys, Ethan welcome to my home. We look forward to you treating our aches and pains doc." He took a good long look around the room and craned his neck to see into the next one before Lance smoothly stepped forward and blocked his view. Ted's smile dimmed but he side stepped around Ethan so he had a direct line of sight to the dining room table we had hauled over and the girls sitting at it.

"Ladies, welcome to my home! We're so happy to have such loveliness brighten our dreary halls!"

Belle and Sasha stared back at him but only gave brief nods. Matty, unaware of the tension in the room bounced in his seat.

"Hi, I'm Matty! Do you have kids I can play with?"

Ted's lost his smile completely for a second before he flashed it back on at Belle.

"Why, what a charming child! Is he yours?"

Again Belle didn't speak but gave another sharp nod pulling Matty's chair closer to hers.

"Well, I do believe there are a few other children around here somewhere but not very many. Such a hard cruel world we now live in that such…fragile children don't survive long."

His tone was sweet and wistful but I took his words as a veiled threat so I moved behind Belle and placed my hand on Matty's shoulder before speaking to him.

"Matty doesn't need to be strong when he has all of us protecting him." My tone wasn't harsh but it wasn't nice either, causing Ted's eyebrows to lift up in amusement.

"Of course, of course, and clearly you've all done such a good job, so far."

My fist and teeth clenched in anger and I would have said more, but I caught Lance's subtle head shake and hard expression that smoothed out instantly when Ted turned to him.

"Well, I'm very glad to see you all settling in. Now, we should go over the expectations and compensations of you all living in my home."

Lance's face went hard, very hard and he took a heavy step towards Ted causing Micky to step forward also with a gleam of anticipation in his eyes.

"That was already negotiated between us! The deal stands!" Lance said in a menacing tone.

Ted waved his hands around the room taking us all in. "Yes, yes, I just want to make sure everyone is on the same page. As discussed, you will hunt and provide much-needed meat for the community as well as be added to the security roster. Your, umm, *partner*, will treat our ailments and attend to any injuries that might surface. What I wish to discuss is what the rest of your group will be doing to contribute to the good of the community? As the saying goes, there's no such thing as a free lunch."

I hate this guy. He's so slick with his proper words and fake tone when it's clear to see he only wants to insult Lance and Ethan's relationship and have the rest of us under his thumb. I try to keep a straight face when Lance crosses his arms and shakes his head.

"Nothing. They will bring nothing to your community. The meat I will bring in and security shifts, as well as my husband's medical skills, will be more than enough to cover their water rations. None of us will need a free lunch from you or your people. As I told you before, this arrangement is for water and nothing else. We will feed ourselves and protect ourselves. If anything, you all are getting more from us than what we are getting. Should we NEED anything more from you, we will negotiate a trade on a case by case basis. Now, with that being cleared up let's definitely be on the same page. If you need me or Ethan then you may send one, just ONE - of your men to come and get us but other than that there is no reason that any of your men should be in this wing of the hotel. Just to be clear, we are NOT your people. We are trading partners only and that was the deal I made with you from the start. If that no longer applies then we can be out of here in twenty-four hours."

All eyes in the room were fixed on Ted's no longer smiling face. All the fake smiles and slickness were gone as he glared at Lance and then like a switch had been flicked they slid back into place.

"Yes, of course, that was the deal we negotiated and I stand by my word. I merely wondered if your people would be interested in participating with the community. I'm sorry if you took it otherwise.

You may want to train your people in your spare time though. There's a chance we will need all the fighters we can get in the coming months. I've recently received information on a possible threat to us. It seems your well wasn't the only one to go dry. There's a large group up at one of the ski resorts and their well has gone dry also. My informant tells me they're considering making a try for the hotel. Unfortunately, they aren't as civilized as us. I've heard they've become rather…primal, if you will. **MY** men won't be able to protect YOUR people if we're attacked. So you should take precautions."

Lance's face was blank but I could see the fury in his eyes as he gave a sharp nod and then walked to the door and held it open for Ted and Mickey. The two men left without another word. The room was silent as we all thought over the implications of what we had just learned until Lance sat back down at the table. He picked up his fork to eat the now cold dinner in front of him before dropping it back down.

"I'm sorry everyone. This was a mistake. I'm sure that bastard knew about the threat before we made our deal. He was looking for more fighters to handle it. We need to start looking for another base. Boys, Ethan and I will have to hold up our end of the deal so that leaves you two. I know you've both been all over this town by now so I need you to start searching again. We need a place that's close to the edge of town that has a well. We can set up the windmills and batteries to work a pump. Start looking on the other side of the highway on the opposite end of town from where our old base was. If we can find a place over there, it will put quite a bit of distance between this hotel and the ski resort. This isn't our fight and I have no plans on making it ours. We need to be very careful and trust no one but the people at this table. We're getting out of here!"

Chapter Seventeen-Skylar

Man I'm tired. The past week has been catching up on butter and cheese making as well as canning vegetables and fruit. I honestly can't think of a good reason why I do it except Dad started us on it and I've just kept to his plan. We don't even need this stuff. There are so many supplies in this place and the barracks next door that Ben and I could live to a hundred and never run out. I guess it just helps to stay busy and feel like I'm accomplishing something. There's a feeling of satisfaction when I see all those pretty filled glass jars all lined up but man, it's a lot of work.

Ben's taken to video games like a fish needs water. I had to put a timer on his playing after the first few days. The kid would play from morning to night if I let him. He needs to stick to his schooling and chores schedule. If the world ever gets back on track out there he's going to need to know things to help rebuild when he grows up. Sometimes I feel completely hopeless about the future but I can't quite give up. Something will change one day and we'll leave this place and be with other people. At least, that's what I keep telling myself.

I'm both excited and terrified for that to happen. My entire social structure has been one small boy, an annoying computer voice and characters in TV shows and movies. I'm so lonely for a real friend that sometimes I make up situations in my head where I have a group of girlfriends and we laugh and tease each other about boys. Sadly, those imaginary girls are all from the TV shows I've watched. What scares me is that I won't know what to say to real people if I ever get the chance. Or even worse, there won't be any people left by the time I leave here.

My mood's taken a downturn these last few days and depression is starting to set in again. It's time for me to take a walk through the dead forest and sit under the broken sky. It's been months since the last time I went out so I go track down Ben. He's right where I thought he'd be, with the cow. Ben treats the cow like a pet every chance he gets. Dad and I had never named the cow but when Ben started to get older he insisted that we name the beast. So cow became Nods. Dumb name, but he picked it because every time he talks to her she nods her head like she's agreeing with him, must be nice to always be right in her eyes. I wonder if she's as lonely for her kind as I am for mine because I swear she gets excited when Ben comes to visit her. She's like a dog with Ben, nodding her head and bumping it into his scrawny chest in affection. She even gives him nasty swipes from her enormous tongue now and again making him giggle madly. Me, I get the cold shoulder and have to watch and dodge for a kick now and then. Not sure what I ever did to her but I'm so not her favorite.

"Hey Benny-boy, chores all done?" He and Nods nod in unison making me roll my eyes but grin too. "I'm going out today so hit the bathroom and grab a snack. You get your choice of movie or game for the next two hours."

He lifts those sky blue eyes that exactly mirror mine up to me, and they're shiny with happiness before they dim.

"Sky, you'll come back right?"

My mouth rounds in an oh. What, where's this coming from?

"Of course I'm coming back! Why would you think I wouldn't be?"

He looks away and rubs Nods velvety nose before answering me.

"Dad went out and never came back so I just wondered if one time you wouldn't come back too."

I'm at a total loss here. Ben was only two and a half when Dad died. I never told him what happened just that he had gone to Heaven to be with Mom so how does he…

"AIRIA!!!"

"I am sorry Skylar, he asked and I am not programmed to lie."

I close my eyes in frustration for a second before kneeling down in front of Ben and taking his hands in mine.

"Listen Bug, I will always come back, ALWAYS! I will never not come back for you! AIRIA watches out for me when I go out. She's always scanning the area out there for anything dangerous and I take my rifle with me as well. I'm very careful so that I will always come back to you. Ok?" He nods his head and so does the big idiot behind him. "Benny, you know you can talk to me, right? You don't have to ask AIRIA, you can ask me anything. I promise I'll always be straight with you. I'll answer any questions you have."

His sad little face scrunches up in determination. "I do have a question. Can I have another boy come for a sleepover?"

My mouth drops open. That so wasn't what I thought he was going to ask. I take a deep breath to tell him about the fate of the world but he just plows ahead.

"Cuz I saw this movie where two buddies have a sleepover and they have gun battles with orange guns that shoot soft bullets and then they build a fort with furniture and blankets and then they eat pizza! Sky, what's pizza?"

I just stare at him and try to keep the tears that are pushing at my eyes from showing. This kid, this poor sad lonely kid will never have a buddy to have Nerf gun battles with. I have to swallow hard to push down the absolute devastation I feel for all that he will never have.

"Benny, Ben, I'm sorry. There are no other boys that live around here to come and play. I wish there was. I really do! I know it's not the same but tonight we **will** have pizza and then we'll make the biggest furniture and blanket fort ever! I know it's not the same and I don't have any of those toy guns but I do know how to make slingshots. We can have an epic battle together."

His face falls for a second but then firms up into a grin. He throws himself at me and wraps his skinny little arms around my waist. "That's ok Sky. You're my best friend! We'll have a great time!"

I try to push a smile onto my face for him but it's so hard when everything is cracking and breaking apart inside of me. I push him gently away from me.

"I don't have to go out. I'm sorry, I didn't know how worried it made you."

He shakes his head. "No, it's ok. You should go. You always come back happier. Really, I'm ok with it. Besides, you said two hours of gaming time! No take backs!"

I know I should stay in with him but I'm swamped with sadness and sliding quickly into despair. I need to get a handle on my emotions and going out is like hitting the reset button so I just nod.

I take a good long look at him before I step into the decontamination room. His whoop of joy at the massive car crash he has in his video game is like a balm to my heart so I step in and the door closes behind me. It doesn't take me very long to suit up. I stopped wearing the containment suit a long time ago. I now wear a heavy white and grey snowsuit that I found with hundreds of identical ones in the barracks. It has a hood that closes tightly over my head and I wear goggles and a filtered mask to keep out the ash and snow that's always blowing around.

"AIRIA, is the perimeter clear?"

"Skylar, nothing is currently detected. Increased cloud cover has made any satellite coverage unavailable. Do you wish for me to monitor Benny-boy's movements?"

"Yes, same as last time."

"Ten-four little buddy. Safe travels!"

I shake my head as I sling my rifle and clip on my communicator. Sometimes I wish I hadn't of told her to search her data banks for slang and casual sayings. It kind of comes out creepy in her stilted mechanical voice.

It doesn't take me long to reach the hunting stand and climb the footholds. I settle down on the stool and let the silence settle over me like a blanket. My eyes are on the boiling grey and black clouds in the sky as I let the memories of sun, laughter and happiness flood over me. I think about

Mom and Dad and the life we had before the bombs. I make up fake memories of how my life would have gone if they had never dropped and I start to find peace. Mom and I are shopping for my first formal dress. I can see every detail of the perfect princess cut gown when it all goes away at AIRIA's words from my waist.

"Skylar, movement detected."

Everything freezes in that instant. All the memories and unfulfilled moments scatter like leaves in the wind and what replaces it is that I promised Ben I'd come back and the image of a bullet hole in my Dad's back. I shoot to my feet and frantically search the forest for movement but don't see anything.

My voice is a harsh whisper when I ask,

"Where, where are they? How many?"

"Skylar, four life forms detected. They have crossed the perimeter south of your location."

South, south is the direction of the door. Can I make it back without them seeing me? If they see me go in then they'll know where we live. Even if they can't get in they'll know where we are.

"AIRIA, can I make it back to the door before they do? Will they see me?"

"Skylar, detection would be seventy-eight percent probable based on their current location and speed to yours."

I can't take the chance. I have to protect Benny and the location of the door. I have to hide. I scan the forest under me for a place to run to before I realise I'm already hidden. They have no reason to look up if I don't give them one, so I drop down onto my belly and rest my cheek against the frozen wood with my rifle lying beside me, ready if I need it. I reach down and mute the volume so AIRIA won't give me away at the wrong moment and then, then I wait.

Chapter Eighteen-Rex

I kick an empty windshield wiper fluid bottle out of my way as I make my way back to the hotel. It's been another disappointing morning of not finding a new base. Marsh and I have been looking for a week now, me in the morning and him in the afternoon. It's frustrating and just getting worse as the days go by and the tension gets thicker. Moving to the hotel was a definite mistake. The girls and Matty never leave the rooms because there always seems to be one of Ted's men hanging around the doors to our wing. It's almost like they're guarding us and every time Marsh and I leave they ask where we're going. They never get an answer from us, making them more and more hostile every day. We've taken to going out of our way and backtracking our route to make sure were not being followed. I'm starting to think there will be a fight when and if we do try and move somewhere else.

I don't think I've ever seen Belle so stressed out before and Sasha is driving me nuts. She's bugging me every day to go with me on my searches. I know she has a crush on me and I should tell her it's never going to happen but it feels like I'd be kicking a puppy if I do. Marsh doesn't help matters the way he's constantly razzing me about it. The rest of the time he's looking at Sasha the way she's looking at me. Uggg! If she'd just turn her head and look at Marsh, she'd see the way he feels about her and maybe leave me alone.

We don't need this crap right now with all that's going on, especially now that Lance is gone for a few days. Ted sent some of his men to scout out the ski resort group to try and find out if they were planning on making a move on the hotel. He insisted that Lance goes with them because he's the only one with a military background.

We've got to get out of that hotel! I can only pray that Marsh has more luck this afternoon. I scan the parking lot of the fast food place I've been meeting Marsh at to switch off and give him the search map we use to not overlap each other but there's no sign of him. I give him ten minutes and then head towards the hotel. The temperature has been dropping for the past hour and my hands feel like ice bricks in my gloves. I'm so sick of this cold, barren world. I miss the sun and how it used to feel on my face.

I'm almost at the hotel parking lot and I keep expecting to see Marsh heading my way but he's nowhere in sight. I'm starting to get a bad feeling about this when I hear the sound of a metal door bang open and then close. That's probably him heading my way now, but my stomach clenches as I pass the last building that blocks my sight of the hotel parking lot and I see Matty running towards me. My body goes colder than it already is but with fear, not the air temperature. I take off in a run to meet him halfway but after only a few strides I'm slowing down in confusion because he's seen me and he's grinning.

"Matty, what are you doing out here?" I yell across the last few yards making his grin slip slightly.

"Rex! Ethan sent for Marsh to go down and help him with a patient so I ran down here to tell you so you wouldn't be waiting for him!"

I take a deep breath and chant to myself, "don't yell, don't yell" but I yell anyways.

"Matty, you're not supposed to BE out here! What were you thinking?"

The kid's grin is gone and a scowl fills his cute face.

"I'm not a BABY! I'm eight and a half, Rex. I can help out sometimes, you know! I was trying to be nice so you weren't waiting around out here in the cold! Jeeze!!!"

I'm mad, but only because I'm terrified something might happen to him so I start nodding.

"I know you want to help and it was nice of you to think of me but listen, Matty, things around here are…umff!"

Out of nowhere, something dark and smelly is thrown over my head and what feels like steel bands clamp down on my arms. It takes me a minute to shake off the dumb confusion of what's happening. Matty's choked off scream of fear is like a dagger stabbing me with part fear, part fury, and I start thrashing with all I've got until a sharp pain in the back of my head takes me away.

My head is pounding like someone's using it for a drum and my stomach is aching like I took a punch to it. I try and stay still cuz I feel like any movement's going to make me puke. What the heck happened? The last thing I remember, I was in the parking lot looking for Marsh when…Matty! My eyes fly open but there's something covering my face and I can't see anything but dim light through a cloth. My hands automatically reach to pull it off but they're stuck behind me tied up with rope. Something shifts against me and I realize someone's leaning against me. I stay as still as possible and listen hard. There, it's muffled through the cloth but I can hear breathing that hitches in silent sobs.

"Matty?" I whisper.

"Shh!" He shushes me and relief floods through me that he's here with me. "Listen, they're fighting."

I try and get my breathing under control cuz it sounds so loud under the fabric and then I can just faintly hear them.

"…such an idiot! He's going to be so pissed! You two were supposed to grab both the teenagers. Now, what are we…" The angry voice fades out and I miss what it says next but I can put two and two together. They were supposed to grab me and Marsh but Matty was out in the lot instead. I strain harder and hear, "take him anyway. Get going and get back here quick." I try and process what's happening through the pain in my head. What do they want with us? They have to know our people will come looking for us, right? Who took us? Is it people from the ski resort or Ted's guys?

Before I can think of anything else I hear footsteps heading towards us and feel Matty tense against me. Seconds later the bag is yanked off my head and I blink to clear my eyes. A snarl of rage slips out of my mouth when I see two of Ted's men standing over us. The men are like night and day. One of them is very tall and broad across the shoulders and he's scowling down at me. The other one is medium height but skinny like he hasn't had a decent meal in years. He's looking nervously at Matty and wringing his hands in worry. The big guy barks out a command at me.

"Get up!"

When I just sit there looking up at him, his lips compress tightly and he reaches down and hauls me to my feet. The other guy is gentler as he helps Matty up. We both have our hands tied behind us making it hard to stay balanced when big guy shoves me forward to start walking. I take a good look around to see where we are as I walk but there's nothing except dead trees and the mountains in front of me. I almost face plant when I crane my head around to check that Matty's behind me and look past him trying to see if anything's familiar. I lose my balance and stagger to the side, coming down hard on one knee. White hot pain shoots through my leg as I feel a sharp rock dig into my bone and I let out a gasp of pain.

"Get up!" growls big guy.

I try and struggle to my feet as involuntary tears start leaking from my eyes but I end up falling on my side. Matty drops to his knees beside me and starts crying when he sees my tears.

"Rex, are you ok? Rex, are you hurt?"

I shake my head at him and try and push the nausea back from the pain in my knee. I have to stay strong for the kid but he surprises me when his face goes all angry and he yells at the two men.

"He can't get up! You have to untie him so he can walk you, you, asshat!"

Even through the pain and fear a jolt of shocked amusement flares up in me. I've never heard Matty swear before and hearing Marsh's and my favorite term come from him is funny. Little guy must find it funny too because he lets out a snort but quickly holds up a hand as big guy takes a menacing step towards him.

"Whoa, Bo! The kid's right about their hands. We're never going to make it back here by dark like this. Let's tie them in front so they can have better balance."

Big guy's name must be Bo. He glares at little guy before grunting and hauling me to my feet. Minutes later, with our hands tied in front of us we start walking again. I shoot a wink at Matty to let him know I'm good and he did a good job. Then focus on the path ahead of us and try not to step down too hard on my throbbing leg. When I had looked back before I fell, I had seen town in the distance and the distinctive peaks behind it so I now know we're headed north. The ski resort that Lance went to scout is west of town so I don't know where we're going. I ask question after question but all I get in return is a few cuffs to the head from Bo, so I just keep my eyes peeled for any chance Matty and I can use to escape.

The first hour passes with us walking on pavement as I plot and scheme with no results. In the second hour, we start walking off road through trees and dead bushes. We're walking uphill and it's getting steeper and harder to walk with our hands tied. The cold is really starting to set in and I'm worried about Matty. This is probably the longest he's been outside since the bombs dropped. We're both dressed in full snowsuits with scarves, toques and gloves but it feels like the cold is deepening by the minute. Winter gear is something we have a lot of, with the town once being a hub for skiers, there was plenty of snow gear to scavenge at the beginning. The only thing that gives me comfort is how little guy keeps helping Matty over rough spots and steadying him when he loses his balance. I still don't know where they're taking us but I can't imagine little guy being so nice to my brother if he plans on hurting him.

By the time we've been walking for two hours my limp is more of a stagger from the damage I did to my knee. I can feel that it's swollen up and it's almost impossible to bend. I'm slowing everyone down and every time Bo has to haul me forward his face gets a little more mean until he's practically dragging me. I let out an agonized roar when he gives a yank on my arm and I step into a depression and wrench my bad leg further. Bo's growling at me as he pulls me to my feet but that leg is done and I collapse as soon as he lets me go.

Matty throws himself down beside me on the ground in panic. "Rex, you ok? What's wrong, Rex?"

I groan and roll over and try and straighten out my damaged leg. My whole body is rock hard tense as it tries to deal with the pain and my teeth are clenched so tightly I can't get words out to calm him down.

Little guy steps in between me and Bo who looks like he going to strangle me with hands he's made into claws and bends down to feel my leg. When he feels just how swollen the knee is under the snow pants he leans back and lets out a tired sigh.

"Well Bo, unless you want to carry this kid over your shoulder, we aren't going any further.

Bo throws his hands up in the air in frustration. "Mickey said we have to go further so their people won't find them! We're still too close!"

All the blood drains from my face at Bo's words causing little guy to look away from me and stand up. This can't be happening. Are these guys really planning on killing a little kid and hiding his body? Matty starts to cry so I pull him closer and loop my tied hands over his head so I can hug him. I don't know what to do! My leg's not going to get us far if we try and run. Even if we do get away, we have no food or water to stay hidden until it heals enough to get back to town. The only thing I can think to do is make Matty run now. Little guy and Bo have stepped away and they're arguing.

"Matty, you have to go. You need to run right now while they're distracted. Try and get back to town and get Marsh and Ethan to come rescue me. You have to be strong and go!"

Matty buries his head deeper against my neck and shakes it hard. His words are muffled, "NO, I'm not leaving you!"

I try and think of the right words to convince him to go and save himself, but a wave of despair washes over me. It's too late. They're done arguing. Bo comes over to us and pulls my arms up so little guy can pull Matty off of me. I'm trying to fight Bo off and get to my brother cuz he's thrashing around bawling in terror but the big guy just pins my arms to me and hauls me up and over his shoulder like I'm as light as a feather. I can't see anything as my face smashes into his rock hard back but he only takes about ten steps before he flips me back over and I hit the frozen ground hard on my butt. The back of my head slams into the frozen bark of a tree making my vision swim. I'm dazed for the few minutes it takes them to tie me to the tree that my head hit, but I can hear Matty's cries right behind me so he must be tied to the same tree on the opposite side. I give my head a shake to try and clear my vision and look down at the rope that's wrapped around me five or six times pinning my arms and body to the tree.

Bo and little guy step around the tree in front of me and I look up at them through tear filled eyes. Bo's just standing there with his arms crossed looking pissed but little guy frowns down at me sadly before he speaks.

"Look, kid, I'm sorry this had to happen but that's just the cards we were all dealt. Hey, I'm no kid killer. I was an accountant before all this, but I've got my own family I have to take care of so I follow orders." He swallows hard and looks away from me for a minute before his tone softens. "It won't hurt. The cold will just make you sleepy. You'll both just go to sleep and all this…hell…we now live in will go away." He shakes his head one more time and turns and walks away.

Bo looks down at me with complete indifference before grunting, "Night-night." Then he walks away too.

I just sit there stunned for a few minutes. This isn't really happening, is it? They're not just going to leave us out here all alone to die, right? It takes a bit for the denial screaming in my head to fade before I accept that we're on our own and I start working on the rope. I rock back and forth and try and bounce up and down to loosen it but the cold has made it rock solid and there's hardly any give to it. I try and ignore Matty's cries every time I jerk my body against the rope. I know it's pulling against him when I move but I can't just sit here and do nothing. His crying gets louder and louder until I just can't take it anymore.

"Matty, hey, hey squirt! Listen, listen to me!" I yell over his cries until he calms down to just hitching sons. "It's ok, buddy. We're going to get out of here. I need you to calm down and work with me. We can do this but we have to work together. Ok, ok, Matty? See if you can slide around the tree a bit. Try and work your way closer to me!"

I feel the ropes bite into my arms and chest as he struggles to move but I know he's not going anywhere with how tight the rope is against us. It doesn't take long for him to give up and his tiny lost voice breaks something in me.

"Rex, I'm cold and, and I'm scared! Why'd those guys leave us here Rex?"

My head drops forward in resignation. I'm not going to win this one. I failed. After so many years of working to survive and keep Matty safe, I've failed and we're going to die. Man, I'm tired. I feel like I've been fighting for so long and maybe now it's time to rest. I lean my head back against the cold bark of the tree and close my eyes to better remember.

"Hey Matty, did I ever tell you that you have Mom's laugh? It was so awesome. When she laughed it came from her whole body and you couldn't not smile it was so great, just like yours. She would have laughed a lot with you."

I paused and remembered how pretty she was. How her green eyes looked just like mine and Matty's and how they sparkled when she was happy. Even after all these years, I miss her so badly. Matty echoes my thoughts when his sad little voice speaks.

"I miss her, Rex, even though I can't remember her or what she looks like. I miss her. I wish she was here."

I swallowed the tears that want to choke me. "She's here Matty! She's been with us every day, watching over us. She…she's waiting for us buddy."

A muffled sob that doesn't come from Matty has my head whipping up and to my right. What was that? Someone, someone's here…where, where did that come from? I'm about to call out, beg whoever's out there to help us when I hear something I haven't heard in seven years. An alarm screams through the forest, an electronic one.

Chapter Nineteen-Skylar

What the frack? Who does that, who leaves a couple of kids tied up in a dead forest to die?

I had lain perfectly still listening to the two men tie those kids up and what they had said in disbelief. Is the world out here so messed up that this is normal? When I hear them start to walk away, I carefully raise myself up onto my elbows and slide my rifle up to my shoulder and take aim at their backs. I focus on the big one and line up my shot. These animals don't deserve to live. I see the hole in Dad's back and wonder if one of these guys was the one who shot him. I give my head a tiny shake, doesn't matter, they're guilty of what they did today. My finger tightens on the trigger as I watch the big guy's back move further and further away until he's out of sight and I slump back down.

I couldn't do it. They deserved to be shot but I just couldn't pull the trigger. It's not my business anyway. I've got Ben to think about. His safety depends on no one knowing we're here. I can't get involved so I just lie there and the minutes tick by as I listen to the little boy's cries. He sounds so young; I can't help but think of Ben crying like that. When the older one starts trying to calm him down and tells him to try and move I nod to myself. They'll get free and go on their way and never know I was here. But then they don't and the older voice starts talking about their Mom who must be dead from the way he talks about her, and my Mom flashes before my eyes. Her face filled with horror and sadness as we drove past those other boys on the road and a sob of grief and loss slips out. She'd want me to help them.

I will!

I'll cut them loose and lead them away from my door and back towards town. I'll walk them in circles so they won't be able to find their way back to my area or tell anyone where they saw me. It's the right thing to do, except just as I'm about to push myself up and stand, the alarm goes off. I slap my gloved hand down onto it and yank it from my belt. I'm just thinking, what the frack? I shut the volume off when AIRIA's voice comes blasting out of it.

"Skylar, volume override. Satellite in range has detected a large front pushing into your region. Sensors indicate a rapid drop in temperature as well as increased levels of radiation. Return to shelter immediately. This looks like a doozy!"

I'm on my feet in seconds. I'm not wearing a containment suit anymore so I'm not protected at all. I have to get back inside. I practically fall from the hunting stand as I slide down the tree. I take two running steps when a voice rings out from behind me causing me to stumble to a stop. I don't turn around, I just stand there thinking. What do I do? I can cut them loose but they'll never survive out here if the temperature drops enough, never mind the radiation. I would be saving them just to let them die.

"Please, help us! My brother's only eight years old!"

It's the older one's voice that rings out and fills me with guilt. The only way to really help them is to take them inside with me. I can't do that! That goes against everything Dad's ever said! I have my own little boy I have to protect!

"Please, he's just a little boy! FRACK!"

I spin around and lunge towards the tree they're tied to. The knife from the sheath strapped to my leg is out in one swift move and held in front of me as I pound towards them. His face is almost completely covered by his toque and scarf but his eyes are visible and they flare wide in fear as he sees the knife in my hand.

"No, NO NO!" he yells as I reach them and swipe.

The rope barely has a nick in it so I drop to my knees and start sawing at it. I don't look at those big green eyes again but they echo in front of my face again and again as I grunt with effort to part the rope. AIRIA's voice blasts out from my waist again.

"Skylar, GPS location shows you are not returning to shelter. Return to shelter immediately. Cutting it a little close, you are!"

There's something surreal about a computer voice trying to mimic Yoda as I'm fighting to save our lives. I'm going to have to rethink this whole humor thing with her.

The rope finally parts and drops away so I surge back to my feet and grab the kid and pull him up with me. His eyes are just as green as the other ones, but they're terrified.

"Get up! We have to run!" I yell at the older boy who's rolling on the ground trying to stand.

"I can't! I busted up my knee!" He yells back at me. I want to screech in frustration but there's no time so I just grab him and with the help of his little brother get him on his feet. I get his arm around my neck and let him use me as a crutch as I lead them both towards the hidden door to my home and safety.

"Who are you?" He grunts out as we hobble along.

"No talking, more walking! We don't have much time!" I huff out, out of breath from practically carrying the guy. I start to hear creaking from the trees behind me as the cold deepens and then a loud crack as branches start to break.

"COME ON!"

I scream in fear and start moving even faster. I have to give the guy credit, even with a busted leg he picks up the pace and a quick glance at his face shows it to be bone white with pain but he doesn't make a sound.

Every second that ticks away is like a gong going off in my head. Will I feel it? Will I feel the radiation or is it already in me slowly killing away all the healthy cells in my body?

There! There's the hidden door. I wrench the fake rock covering from the keypad so hard that one of the hinges snaps leaving it dangling open. Don't care, code in, door whooshing open to the side and I'm shoving him off my shoulder and into the airlock before grabbing the kid and pulling him in with me.

"AIRIA, CLOSE THE DOOR!"

The door slides shut and I drop to my knees panting for breath. The light changes to red causing the boys to gasp, but I don't bother looking up. I just wait for the scan and the results.

"Scanning for radiation and other environmental impurities. Scan complete. Sterilization commencing."

The light changes again and the kid gives a short scream but again I just ignore them and wait and wait until my head is going to burst and I scream out, "AIRIA!" causing the two boys to flinch and slide further away from me.

"Skylar, trace amounts of radiation found to be at non-life threatening levels. Recommend removal and disposal of outer clothing."

The breath I had been holding gushes out of me and I start unwinding my scarf and unzipping my snow suit. The boys just sit huddled against the wall staring at me so I bark out at them.

"OFF, take your snowsuits off!" When they don't move I yell, "NOW or you can go back out there!" Fear has made me mean but they start moving so I just look away in shame.

I get all my outer gear off but keep my gloves on until I gather all the clothing from the floor and stuff it into a compartment in one of the walls, then throw my gloves in last. I take a deep breath and think before turning around to face the two boys I've broken every rule for.

"AIRIA, Scan again."

"Repeating scan... Scan complete. Sterilization commencing."

I wait with my back to them until the UV light changes back to the fluorescents and AIRIA speaks.

"Sterilization completed. Recommend...thorough scrubbing with soap and water for your two guests. Voice imprint of guest will be required for access."

I let my shoulders slump. Ok, we made it, we're fine. I slowly turn around and take a good look at the two people I risked my life and maybe Ben's for.

The older one is at least a half a foot taller than me with dark brown matted curly hair. He's probably close to my age and his green eyes are studying me just as closely as I'm studying him, making me uncomfortable, so I look away to his little brother. And here's cuteness in a can. He has the same hair and eyes as the older one but he tries to give me a tentative smile, popping a pair of matching dimples in his cheeks. My lips tug up in return and I think he's around the same age as Benny.

BENNY! My eyes flash back to the older boy.

"Are you sick? Are either of you sick with anything?" My heart starts pounding again in fear. Benny, who was born inside away from the regular and scary germs and diseases that the real world holds, might not be able to handle germs. He's never even had a cold. Does he have an immune system? Can I take that chance?

"We're not sick! There's nothing wrong with us!" the older boy says sharply before asking, "Who are you? What is this place? Who's that talking through the speakers?"

I totally ignore him cuz before I can even think about letting them in, I need to know that Benny will be safe.

"AIRIA, will Benny get sick if he's exposed to these people?"

"Skylar, bodily fluids will need to be tested to determine if they are carrying any internal viruses that may infect Benjamin or you. Benjamin has been vaccinated for all childhood illnesses but will require the booster shots that are suggested when a child reaches the age of five. Shots, you did not give!"

What? What shots? Dad must have given them to Benny when he was a baby. I'll have to find out where they are and give him the booster shots, but right now I have to deal with these two.

"Screw you! We're not giving you any bodily fluids!" The older kid snarls while he takes a step towards me.

The gun I have in a holster against my back is in my hand in a split second and pointing at his face. I always wear it under my snowsuit when I go out in case my rifle jams. I'm always prepared out there after what happened to Dad.

His face freezes and he takes a slow step back pulling his brother behind him. The little, scared face that peeks out reminds me that they're just as scared and confused as I am, so I holster it and hold up my hand asking for patience.

"Ok, let's just take a breath here. My name is Skylar. I have a little brother inside that's just a bit younger than yours. He's never been outside in his whole life and he's never seen anyone but me and my dad. He's never even had a cold. I'm just worried his immune system won't be able to handle anything you might bring in."

He looks down at his brother and then back at me before nodding. "Ok, I get that. I'm always worried about my brother too. But, what is this place and what's inside? I mean, thank you for saving us and for getting us to shelter but where are we?"

I let out a sigh. There's no point standing out here while I try to explain so instead of answering, I ask, "What's your names?"

"I'm Rex, this is Matty."

I nod and give a half smile.

"Hi, Rex. Hi, Matty. Matty, can you say your name and how old you are, please."

Rex gives me a weird look but nudges his brother out from behind his back. Matty looks at me with uncertain eyes before looking all around the room and then up at the ceiling.

"I'm Matty and I'm eight and a half years old. Where's the other lady hiding?"

It takes me a second to get what he's asking but then I understand that this boy would have been a baby when the bombs dropped and he probably has never seen or heard of a computer before. I can't help but smile, he's so cute.

"AIRIA, did you voice imprint Rex and Matty? Authorization level Red."

"Skylar, voice imprints and authorization recorded. Are you going to stand there all day? Benjamin is aware that you are in the airlock and is voicing his displeasure that you are taking so long and ignoring his calls."

I frown and look around for my communicator that I dropped when I took off my gear. I spot it in the corner and snag it. Right, the volume is still muted from before. AIRIA must have just overridden it for the alarm. I turn the volume up and sure enough, there's Benny squawking my name over and over. I hit the send button and try not to laugh when Matty's eyes go huge at the sound of Ben's voice.

"Benny, chill out! I'm here and I brought you a surprise!"

I shut the power off and drop it into its charger in the closet before he can start calling me again and turn to the two boys.

"So, we should go in. Umm, welcome to my home."

I palm the door control and watch as their eyes go even wider in surprise when it opens. I walk through but have to turn and beckon them forward when they just stand there. They both jump a bit when the door slides closed behind them.

Benny's eyes are glued to the game he's playing on the TV but he's talking nonstop.

"Hey, Sky! What took you so long? You were out there forever this time! What kind of surprise did you bring me? Is it something from storage? Is it new food or a new game?"

When I don't answer him, he shoots a look over his shoulder before looking back to his game and then the controller drops from his hands and he's turning in slow motion with his sweet lips in a perfect circle of surprise. His eyes hesitate on Rex before sliding down to Matty. His eyes are huge with astonishment when he whispers.

"Sky, you brought me a boy?" Then increasingly louder, "You brought me a BOY? Oh man, oh, man, whoo hoo! You brought me a boy!" He scoots around the back of the couch and comes charging over to us with the biggest grin on his face.

"Hi, hi, I'm Ben, what's your name? Do you like video games? Do you want to help me build a fort? Sky's gonna make pizza! Do you want to see my room? Do you…"

As scared as I was a few minutes ago is how happy I am right now seeing the pure joy on my baby's face. I look over at Rex who is also smiling and wham! He's got a pair of dimples to match his brother's except his aren't cute, his are like weapons that punch me straight in the gut. Holy crap! Where did that come from? Forget it, deal with motor mouth.

"Whoa, Benny! Back off a bit. Take a breath little man. This is Matty and this is his brother Rex. They're going to be here for a little while until some bad weather clears up outside."

Benny beams all the joy in the world from his eyes and claps his hands in excitement. "Can Matty have a sleep over? We can build a fort and sleep in it and use flashlights and…what's that smell?"

My face goes red in embarrassment. The boys don't really smell bad, but they don't smell great either. I try and stop Benny from saying anything else by cutting him off.

"Oh, I'm all sweaty! We had to run for the door and Rex hurt his leg so I had to help him so I really worked up a sweat! We should all have showers and get cleaned up before you guys go play and I start on dinner. Rex, can you manage on that leg? I have some fresh clothing I can lend you guys if you want?"

I bite my lip and feel like an idiot when Rex's face turns bright red and he looks away with a nod but it's too late now and really they probably would feel better if they got cleaned up. I discreetly look their clothes over as they hobble to the bathroom that I pointed out. They're not dirty or stained but they are worn and have a few patches here and there.

All of a sudden I'm nervous. I don't know why except no one's ever been here before and I don't quite know what to do. Before the bathroom door closes behind them I catch a glimpse of Rex's stormy face. Oh my God, I totally embarrassed him! Oh no, what if he's mad at me? What if he doesn't like…WAIT! What the frack was that? I start to shake my head, how did that happen? Two minutes with a boy that has amazing dimples and I start acting like, like…urrrggg…like a teenage girl!

Chapter Twenty-Rex

I close the bathroom door and lean my back against it to take the weight off of my leg. My head is spinning with all that's happened in a very short time. I need a minute to process the fact that Matty and I were just about to die in that dead forest, and now we're in a place that was never touched by the bombs. Right, let's not forget the gorgeous superhero girl that swooped in and rescued us and took us to her bat cave. I need a minute, but Matty's caught the other boy's question bug and he's rattling them off a mile a minute, so I push off the door and hop over to the shower stall that stands beside a huge soaker tub.

"Why do all the lights work here? What's a shower? Is it like a taking a bath? Where do we get the buckets of water? Will the water be cold or can we heat it up, cuz I'm still really cold. Who's that boy, Ben? Can he be my friend? Rex, Rex, what's pizza?"

I pull the glass shower door open and take a deep breath of the amazing smell that comes out. There are bottles of shampoo and body wash in there that smell like flowers so I'm guessing that the water will work. I turn away from the stall and look Matty over then take a sniff. Ok, compared to the flower smell, we don't smell that great. Before we moved to the hotel and the well still had water we would all bathe every few days, but we had to haul buckets at the hotel so none of us had done more than a quick washcloth wipe down since we moved. This was going to be heaven, especially if the water was hot.

"Alright squirt, pipe down and strip. There'll be plenty of time to talk about all that after we get clean."

Once I finally got the kid undressed and the water flowing, it was hard not to laugh at the way he squealed when he stepped in under the spray. You'd think he had never been wet before the way he kept dancing around in there. By the time I got him soaped from head to toe and rinsed off, I was half- drenched myself, and the throbbing in my leg was almost unbearable. I had to sit on the edge of the big tub with it off to the side while I wrapped Matty up in a massive, soft, sweet smelling towel.

Everything hurt. There was a knot on the back of my head where Bo hit me the first time and more tenderness where I had hit my head against the tree. I'm pretty sure my tailbone was bruised badly when he dropped me down from his shoulder as well. I wanted so badly to just lie on the floor of that shower and let the hot water beat against my aches and pains. A soft knock on the door brought me out of my pity party and back on my one good leg as I used the vanity to steady my progress to the door and open it.

Skylar was standing on the other side with a hesitant smile and a bundle of clothing balanced on one arm. In the other, she held out a bottle of water and a pill bottle.

"Hey, I um, thought you might need something to help get the swelling down on your leg. They're expired, but AIRIA says that they will still work if you take three or four of them. I also brought you both some fresh clothes. Ben's stuff should fit your brother and my dad's clothes might be a bit big, but they should work."

I take the bundle of clothes from her and set them on the counter before taking the water and pills. I have so many questions I want to ask her, like how she has hot water, electricity and a talking computer. Not to mention, where is her dad? I just nod and mumble a quick thanks. I'm so exhausted and sore that I still need time to figure some stuff out before we talk. She nods back and starts to turn away before quickly turning back.

"Oh, you could take a bath instead of a shower if it'd be easier on that leg. It might help to soak it. I have ice too, for after. Um, and it's ok if Matty wanted to come out if he's done. Ben's climbing the walls in excitement."

She's different now than the hard girl that yanked me to my feet and put a gun in my face, softer and a bit uncertain so I give her a better smile.

"Thanks, I think a soak in hot water would help a lot but I'd like Matty to stay with me until I'm done. We won't be too long."

She nods her head quickly. "Right, right, of course. He should wait for you! Umm, I know you don't know me and I was a little harsh at first but he's safe. I mean, both of you are safe, here, with Ben and I." She rolls her eyes and huffs a curl of hair away from her face. "Right, ok, I'll um, see you in a bit then."

I close the door with a grin. I make her nervous which is ok because she sort of makes me nervous too. There are other girls at the hotel and Sasha's always under my feet but this girl's something else. She's so tough the way she handled herself outside and in the airlock and the way she looks, wow, just wow. Her hair is seriously like waves of gold that flow down her back and her eyes are a reminder of the prettiest summer sky. I don't know her or her story but I'm really interested to find out.

After downing four of the ibuprofen pills, I get Matty dried off and dressed while the big tub fills with steaming water. Matty's still gabbing on about what he thinks video games and pizza are, so I just tune him out and slowly lower myself into the tub. I can't stop the whimper that slips out - half pain, half pleasure. I settle in after I wash my hair and soap up, close my eyes and try and make sense of everything that's happened today. I hear Matty talking and talking until I don't.

A cold shiver brings me back and I shoot up in the tub. The water's gone cold. I must have fallen asleep but even worse, Matty's gone! I curse under my breath and pull the plug before getting out. I feel better, the chill is gone from my bones and the pain in my knee is softer unless I try and put any weight on it. Thankfully, the clothes Skylar gave me are soft, easy on baggy sweats. I give my hair a good rub with the towel to get it close to dry and then try and clean up the bathroom as best I can. Judging by the clean smell of the towels, she's got a washing machine that works somewhere in this magic mountain, so I gather up all of our clothes and used towels before opening the door. One step out of the room and I walk smack into a smell from my childhood. I'm not ashamed of the tears that prick at my eyes. I mean, come on…PIZZA? I didn't really believe the kid when he said pizza!

Chapter Twenty One-Skylar

I can't help but smile at the laughter coming from the two boys as Ben tries to teach his new friend the ins and outs of car racing in video games. I've seen Benny happy before but not this kind of vibrating, bust out of his skin in excitement happy. He needs this. He needs to be around other kids and if I'm honest, maybe I do too. An hour ago I was going to walk away and leave these boys to die. What kind of person does that make me? Yes, Ben and I have been safe here all alone but was that really enough? I'm starting to change my mind again. Maybe Dad was wrong. Maybe he should have tried to help people. I can't believe everyone out there is as bad as the people who shot him in the back and left these two boys in the woods to die.

I glance over at the bathroom door and wonder if I should check on Rex. He's been in there a long time. His brother said to leave him but…no, he'll be out when he's ready.

I look down critically at the finished results of my first pizza making. Mom and I used to make pizza all the time before she died but it was always from a kit. This looks like pizza but I'm sure it won't taste like the ones we used to make or have delivered. I made the crust from scratch following AIRIA's instruction and it looks like pre-baked pizza dough, but the pepperoni and cheese are another matter. We have so much meat in the walk-in freezers that I'll never be able to eat it all. It still tastes like the meat I remember, just less somehow. AIRIA says it's because it's been frozen for so long. She says it's safe to eat but that the longer it's frozen the less flavor it will have. It's better than the dehydrated meats we have in storage though. The pepperoni still tastes like pepperoni but it's like chewing on flavored leather after it was dried, smoked and then frozen. The mozzarella is homemade so it's softer and doesn't shred up as well as the cheese we used to buy in a grocery store and of course, the sauce was made from tomatoes we've grown and then canned with spices. So, it's definitely NOT delivery pizza but I hope Benny and the other two will still like it.

I pop the two pizzas in the oven and start making a dressing for the salad I plan on serving and glance at the bathroom door again. I have so many questions to ask him about what it's like out there. Funny, he's barely spoken ten words since I let them in here but I just have a feeling he's a good guy. I guess it's because of the way he takes care of his little brother. I've just finished setting the table for the first time in forever, after clearing mounds of schoolwork off of it, when the bathroom door finally opens. I spin around and see a very clean looking Rex standing just outside of it with his arms filled with dirty clothes and towels and a really weird look on his face.

"Oh, hi! There's a hamper right there beside the door you can put those in and I got you a crutch. I thought it might help you get around until your leg feels better. Dinner's almost ready if we can peel the boys away from their game."

I stand there like an idiot just watching him nod, dump the clothes and grab the crutch that I had left. Is this guy ever going to speak? The timer on the oven goes off so I turn away and get busy getting the food to the table after I call Benny and Matty over. Once we're all seated and the food is ready to be dished up, we all just sit in nervous silence for a few minutes before Rex speaks.

"Is your dad going to be joining us?"

I feel my face pale and look quickly to Ben before shaking my head.

"He… he died a few years ago. It's just Ben and me here."

Rex looks down at his plate before meeting my eyes again. "I'm sorry, I didn't know. I umm…" He looks around the table and picks up his glass. "You have milk? Is this real milk? Like from a cow or is it from a box?"

I'm grateful he's changed the subject. I don't want to talk about Dad and what happened in front of Benny.

"Oh, yes it's real! We have a cow."

He points to the salad with a raised eyebrow and I give a little laugh.

"Yes, we have a garden too."

Rex shakes his head in wonder. "This, this is incredible. Can I ask how? How do you have all this stuff? Running water, electricity, all this…food?"

I open my mouth to tell him about my dad and Uncle Bill but I stop. Dad's words ring through my head about how if anyone knew where we were and everything we have that we'd be overrun. Rex already knows the area where we live and now he knows we have lots of supplies so what's the point in keeping everything else from him?

He sees my hesitation and waves his question away.

"Hey, it's ok! You don't have to tell me anything. We're just grateful that you took us in and fed us. This pizza is so awesome! And hey, Matty, how great is it to not to have to eat sprouts for dinner?"

Matty's mouth is full of cheese and pepperoni so he just makes his eyes really big and nods exaggeratedly in agreement.

I'm thankful he doesn't press his question so I try and joke.

"Oh well, we have sprouts if you want some! They're very good for you, you know and they grow super-fast!"

Both Matty and Rex groan.

"Yeah, that's been one of our main food sources for the past seven years! Belle puts them in everything we eat. It got old really fast! This meal? This is like a dream meal, even the salad is amazing. I never thought I would ever eat pizza again."

He's opened the door to how they live outside and I can't resist asking about it.

"So, you live with other people? Do you have more family out there? Do you live near here?"

His face clouds over and he shoots a glance at his brother but he and Ben are whispering to each other so Rex turns back to me and lowers his voice.

"We do have people, a family, sort of. Our Mom died on the first day but a lady, Belle, and her daughter found us and looked after us. We joined up with another family shortly after that and have been together ever since. They're really good people, we're a family now." I can tell he wants to say more but another look over at the boys stops him.

There's not much left of the pizza and the boys are fidgeting so I send them back to their game and start clearing the table. Rex gets up and helps me carry plates to the sink and I can't help but remember how normal this used to be, Mom and Dad cleaning up dinner while I flopped in front of the TV in our old life. I wash, and while Rex dries, he continues what we were talking about at the table.

"We had a really great place we lived in all together for seven years but a few weeks ago we lost our water source. We were forced to move in with another group that still has water in one of the big hotels. Most of the people there are nice enough but the people who run it aren't. The leader of that group, Ted, was the one who ordered his men to get rid of Matty and me but from what I overheard it wasn't supposed to be Matty but me and Marsh. Matty was just in the wrong place at the wrong time."

I hand him the washed pizza pan to dry with a frown.

"Why? Why does this guy want you and uh, Marsh, dead?"

"Yeah, Marsh is part of the other family we joined up with. He's the same age as me. He has two dad's named Lance and Ethan. Lance is sort of our leader because he was in the military and knows all kinds of survival stuff. Ethan's a doctor but he was also in the military as a field surgeon. Those guys saved our lives. There's no way me, Matty, Sasha and Belle would have lasted so long without them.

Anyways, I don't know why for sure but Ted doesn't like us very much. Lance and Ethan being gay was strike one in his and his men's book, then Lance wouldn't bow down to Ted and be under his control which is strike two. I think the biggest reason though is that he wants our supplies and the girls. He made a few…suggestions about them that weren't very pleasant and Lance shut him down hard. We've done pretty good by apocalypse standards and Ted wants the supplies we have, probably more than he wants the girls. He sent Lance off on a scouting mission with some of his men a few days ago, so I'm really worried he'll try and take him out too."

Rex looks over at the boys when they whoop and yell at something from the game before turning back to me with a hard look.

"Listen, I need a huge favor from you. I know you don't know me or know what kind of guy I am but you seem like a really decent person and I'm getting pretty worried. I need to get back to town to the rest of my family and warn them or try and help them if Ted's already made his move. I need, no, I'm begging you to keep Matty here where he'll be safe. I know it's a lot to ask, with us just meeting, but I can't have him in danger!"

Chapter Twenty Two-Rex

She just stands there with wet soapy hands staring down into the sink. This is such a huge risk, probably the biggest risk I've ever taken in my life but I don't know what to do. I have to get back to town and I can't risk Matty's life. I know in my gut I can trust this girl with my brother. She would keep him in here safe with her own brother. I also know that if something happens to me and the rest of my group out there she wouldn't turn Matty out. He'd have a safe place to grow up. I'm about to try and make my case again when she finally looks up at me.

"When, when do you want to leave?"

The breath whooshes out of me in relief.

"As soon as I can."

"Alright, let's think about this then. It's dark outside right now so you'd have to wait at least until morning, right?"

I groan but nod in agreement so she goes on.

"The terrain between here and the closest road is pretty rough. How's that leg feeling? Do you think you'll be able to make it very far on it?"

I look away at from those blue eyes and grit my teeth. "I can suck it up. My family's lives depend on it."

She nods her head. "Fair enough, if it was my family I'd do the same thing. We can wrap that knee up tight to help and dose you up on pain killers. I'm fine with Matty staying here but I have to ask, what's to stop you from bringing a bunch of people back here and taking over this place for yourselves?"

Her eyes are dead cold serious. My mouth opens and then closes with a snap. She's right, why should she trust me? She doesn't know me and this is a really sweet set up that many people would kill for. I don't have a good reason to give her so I just shake my head.

"There **is** no reason I can give you to trust me."

She looks hard into my eyes and then looks over at the boys before speaking.

"My dad, he said never to trust anyone, never help anyone, always stay hidden. I've been alone with Ben since I was thirteen. I…I think maybe he was wrong. I'll keep Matty safe and I'll trust you not to bring an angry horde with you when you come back for him."

My shoulders slump in relief. I owe this girl mine and Matty's lives. I would never do anything to put her and her brother in danger. She gives me a brief smile before addressing the computer.

"AIRIA, what's happening with that radiation cloud outside?"

"Skylar, it's a doozy!"

I must have made a weird face because her face goes red and she snaps out, "AIRIA, humor OFF!"

"Skylar Ross, radiation cloud is expected to clear region within three days."

My fists clench in frustration. Three days? I can't wait three days to get back to town. Anything can happen in three days! And then something else wipes that away.

"Ross, your name is Skylar Ross?"

She gives me a confused look but nods her head yes.

"Your father, was his name Daniel Ross?"

The confusion leaves her face and terror replaces it for a fraction of a second before they go hard and cold. She reaches behind her back but her hand comes out empty so she lunges for the knife block on the counter and pulls a huge butcher knife out and points it at me.

Chapter Twenty Three-Skylar

Where the frack is my gun, dammit! My heart feels like it's going to pound right out of my chest. I let my guard down. Stupid lonely girl falls for stupid dimple boy. The only way this guy knows my dad is if he talked to him before he died. Like just before he put a bullet in his back.

"You killed my father!" I hiss at him.

His face goes blank in shock and he jumps back a bit while thrashing his head back and forth in denial.

"NO, no way! Your Dad's my hero. He's one of the main reasons Matty and I are still alive! He helped us!"

I shake my head. No way, not my Dad.

"My Dad never helped anyone, ever. It was one of his biggest rules! Now tell me how you know him!" My voice has risen and out of the corner of my eye, I see the boys have stopped playing the game and are staring at me in fear.

"Wait, he did. He did help us. That first day when all the bombs dropped and everything died, he, you, all drove past us. Matty and I were standing by the side of the road all alone and your truck passed us. You waved at me! He came back! He came back later and you and the lady in the front weren't in the truck anymore. He came back and he took me, Matty, Sasha and Belle to town to the superstore there. He told us what happened to the world and he helped us get supplies and barricade a room where we'd be safe. Skylar, your Dad saved our lives. The last time I saw him I was eleven and Matty was one and a half. I didn't kill your Dad!"

Oh God, oh God, if he's telling the truth that means I killed her. It's my fault she's dead. I treated him so badly for not helping all those people that he must of went back out of guilt. He was with them when the baby came and Mom died. It's all my fault he wasn't back here to save her. The knife drops from my trembling hand and clatters into the sink. I'm shaking so bad and then I'm doubling over with the pain of it as sobs burst out of me. I'm so lost in my guilt and grief that I don't even resist when Rex scoops me up and carries me to my room even though it must kill his hurt knee. I think I hear him calming the boys but I just don't care. My Dad, he helped people. He wasn't a coward but he wasn't there to save Mom because of it.

Rex sets me on the bed and grabs a tissue box from the dresser. He doesn't say anything as I cry out seven years of repressed pain. He just kneels in front of me on the floor with his bad leg to the side and his hand on my shoulder until my crying slows down to hitching breaths. Then he hands me the tissues and waits for me to blow my nose before simply saying,

"Tell me."

It all pours out, that day, that awful day we ran from our home and our lives. Passing all those people without stopping to help any of them and the harsh words I said to my dad about it. Coming here and learning how the world died and then Mom having the baby with just me there and her dying because I made him feel guilty so he went back to help. I told him about finding Dad's body and hating the people in the town and wishing they would all die too.

"All those years, I thought he was a coward to never help anyone but he did help and because of that Mom died. It's my fault and **he** died and now I can never say I'm sorry!"

Rex just stared at me with sad eyes before he took my hands in his.

"I was eleven that day. We were moving from B.C. to Alberta and I was so mad at my mom. For days before that, I gave her the silent treatment and when I did talk to her it was always in a snippy tone. I blamed her for ruining my life, making me move away from all my friends and my dad. She, she died in a car crash when everything stopped working. I never got to say I'm sorry either. But here's the thing Skylar, we were just kids. There was nothing we could have done to save them and I know, I know they knew we loved them and they also know that we're sorry."

I sniffed my nose and cleared my throat but my voice still came out a croak.

"How? How do you know that?"

He smiled in understanding.

"Because, you and me, we're parents. When Ben is having a bad day and being mean to you, do you know he still loves you? And when he does something wrong, do you know he'll be sorry for it?"

I close my eyes and nod yes. I do know those things because I am Ben's Mom even though I'm his sister.

"That's right. We know what it is to be a parent. Skylar, it wasn't your fault your mom died. Even if your dad had been here, it probably wouldn't have mattered."

I shake my head in denial so he tightens his grip on my hands.

"Was your dad a doctor or a surgeon? Could he have operated and saved your mom?"

I feel the weight that's always been pressing against me lift off at his words. I know he's right but it'll take a long time for that guilt and grief to totally fade away and heal after so many years of carrying it. I catch movement out of the corner of my eyes and turn my head to see two sad little boys with tears running down their faces and arms around each other standing in the door. Ben's mouth is trembling with tears when he speaks to us.

"None of us have moms and dads. We should all be a family so we won't be alone anymore."

Matty nods his head in agreement so I turn to look at Rex and take a deep breath and let a small smile lift my lips.

"I think he's right. I think that everybody who's left should be together and help each other so no one's alone anymore but first, first we should have ice cream!"

I try and lighten the mood because I can't stand seeing those small boys so sad. There will be time to think about the spark of an idea Ben's words have given me later.

Chapter Twenty Four-Rex

I steal a glance over at Skylar at the other end of the couch when the credits start rolling on the movie we just watched. The boys are curled up asleep, half under the coffee table, covered in a blanket we turned into a small fort. The giant robots that can change into vehicles weren't able to compete with the excitement of the day and the boys only made it through half the movie. She's staring at the TV but I don't think she's really seeing it.

I'm in awe of this girl. She's so strong from all that she's had to deal with alone, but she's broken too. I'd say everyone who's survived this long is broken in some way but most people haven't had to be all alone dealing with it. I think she's embarrassed by her breakdown cuz she's barely looked at me since then. As soon as we left the bedroom she's acted like nothing happened, getting bowls of homemade ice cream sprinkled with chocolate chips for everyone and setting up the fort for the boys. I wish I knew what she was thinking. She looks so lost that I can't help but try and bring her back.

"I just want to thank you again for everything you've done for us. Today's been one of the worst and best days. For a while there, I thought Matty and I were going to die. What you did for us, saving us from that is not something I will ever forget. Taking us into your home and trusting me, I can't ever repay you for that." When she still doesn't respond I continue. "Seeing Matty with your brother and him getting to do all the things that normal little boys used to do is such a gift. He's missed out on so much since the bombs dropped. Every day has been just focusing on survival so a lot of the joy and fun we used to take for granted is lacking in his life. You gave him, well both of us, a gift by letting us in here." She's still just staring at the now black screen of the TV and I'm feeling pretty awkward so I try a last resort to engage her. "I'm pretty much throwing myself at your feet here in worship for the most amazing thing you gave me…chocolate! Do you know how long it's been since I've had chocolate of any kind? But chocolate chips and ice cream? Seriously, my life now belongs to you!"

Finally, she turns her head slowly to look at me in confusion before she registers the silly grin on my face. Her eyes clear and a grin tugs at her own lips. I hope she's going to open up again but she just shakes her head in amusement and pushes off the couch to her feet and looks down at the two sleeping boys.

"Should we carry them into the bedroom or just let them camp out here for the night? If we leave them here, you can sleep in Ben's room."

I grab the crutch from the floor and clumsily get to my feet so I can see the boys. They look so cute curled up together that I just shake my head.

"Nah, let's leave them be."

We just sort of stare at each other for a minute before she nods, "Ok, well, have a good sleep. I hope your leg feels better tomorrow."

I just stand and watch her go feeling uncertain until she reaches her door where she turns and gives me one last look.

"Rex… I'm glad you're here. I mean, I'm glad I could help you and Matty and I really hope that I can trust you. Goodnight."

Chapter Twenty Five-Skylar

"Dude, that's EPIC!" wakes me the next morning. I just lay there completely exhausted from a night of tossing and turning. I spent most of the night rehashing every memory I have of Dad after the bombs fell and what I should do going forward. Having Rex and Matty here with us has made so many things I thought were set in stone, shift. Ben's soaking up Matty's companionship like a dying plant does water and having Rex here has made me feel things I've never felt before. I can't help but think maybe Ben and I were only half alive all this time and maybe, maybe it's time to finally really start living.

Peels of little boy laughter ring out from the living room making me smile. Yeah, maybe it's time to find some real happiness instead of just existing. I roll out of bed and quickly get dressed. I'm sure there are some hungry boys that could use breakfast. As soon as I open my door I freeze. The distinct smell of frying bacon hits me and I try and process how I feel about Rex helping himself to my kitchen. On the one hand, what the frack, who does he think he is? On the other hand, the last time anyone made me breakfast was when Dad was alive.

He has his back to me at the stove when he calls out, "Alright you speed demons, game's over! Go wash up and get over here before I gobble up all this food and make you eat broccoli for breakfast!"

I grin as the boys make fake gagging sounds and race to the bathroom to wash their hands. Rex turns with a plate full of bacon and eggs on it and almost fumbles it as he sees me standing at my door. He breaks out those big dimples and my stomach flutters.

"Hey, good morning! I hope you don't mind me invading your kitchen. Those two were grumbling for food the minute their eyes opened. I just thought I could maybe give you a break and let you sleep in for a little while longer and, well, you have BACON!"

I can't help but smile. This guy, he's, well, hmmm…

"No, seriously! I gave you my life for that chocolate, but bacon? Skylar, you now own my soul."

He says it so seriously that the laughter just explodes out of me. I have to wipe away the tears I'm laughing so hard. Ben and Matty come racing out of the bathroom but Ben comes sliding to a stop and gives me a weird look.

"Sky, are you ok?" He asks me in a worried voice.

"Yes, of course. Why?"

He shakes his head and starts to grin.

"Cuz I've never heard you laugh like that before!"

I look over at Rex and smile.

"Yeah, well I think there'll be a lot more reason to laugh from now on." He gives me a small nod of agreement so I let my heart lead. "After breakfast, I have some things I'd like to show you."

We dig into bacon, eggs and toast with lots of laughter but I can see the strain around Rex's eyes. I guess correctly that he's worried about the rest of his people down in the town so I ask AIRIA about the weather but it's still too dangerous to go out there. Once we've cleaned the table of dishes, Ben and I give Rex and Matty a tour of the cavern and all the back rooms. They ooh and awe over the animals and garden and all the other rooms, but it's the barracks that affects Rex the most.

He just stands there looking at the huge room with a grim face as I tell him about the storage containers of supplies I found. The boys are running wild around the hundreds of bunks in a game of tag as I wait to see what he'll say. As his silence drags on, I take a deep breath and ask him.

"How many people, good people, are in that town that could come and live here?"

His head whips towards me with a face filled with hope. His crutch clatters to the ground as he wraps his arms around my waist and spins me around with a shouted "Whoop!" He only makes it one spin before he lets me go and bends over to clutch at his swollen knee.

I'm laughing too hard to scold him properly as I retrieve his crutch and get it under his arm again. Once he has his balance, he rubs his hand over his eyes before he pierces me with a serious look from those green eyes.

"Do you mean it? You have to really mean it Sky. This is a really big deal. There's over a hundred people in the hotel alone. There are other groups spread out in the town as well that we've traded with in the past. That's a huge step for you, to go from living all alone with Ben to sharing your home with a whole town."

I look out over the rows of empty bunks and nod without looking at him.

"He should have done it from the start. This place can hold over a thousand people and yet here it is empty except for two. How many people died because he kept it for us?" I turn and look at him. "It's time to start rebuilding. It's time to start fixing what was broken."

A warm glow starts to fill me when Rex takes my hand in his and we look out over the empty barracks that ring out with boyish laughter. Yes, it's time to start living again.

Chapter Twenty Six-Rex

I'm so excited and so scared all at the same time. Skylar and I spend the rest of the day making plans to move the survivors from the town up to the barracks. We'll need a lot of help to co-ordinate everything but first Ted and his men need to be dealt with. I need to get down there and find out what's happening at the hotel. Every moment I stay is a double edge knife. Being here with Skylar is amazing. In only one day I've already come to have feelings for her and I just want to stay here with her and the boys forever but my people, my family, are in danger and I'm going crazy with worry.

Skylar wants to come with me to the hotel but we both agree that she needs to stay here with the boys to keep them safe and start pulling supplies from the storage containers to get the barracks ready. She finally agreed to stay with the condition that she's coming after me if I'm not back in three days. She's so fierce in some ways and so soft in others. I've fallen for this girl and I can't wait to see what our future holds.

"Rex, are you listening to me? If I can't go with you then I can at least give you an advantage. Those guys were going to let you and Matty die. You can't go back and think you'll just be able to talk them into being nice! You need to be prepared to use force if they've taken or hurt your people."

She gives me a hard look so I nod. She's right, things are going to be ugly so I need to be ready to do what I have to.

"Good, do you know how to shoot a gun?"

I blow out a breath. "No, but I do know how to use a bow."

She looks up at the ceiling and thinks for a second before giving me a grim smile.

"Ok, I don't think that'll be enough but it's a start. Follow me."

We leave the boys playing in a sandbox beside a small stream cut into the rock floor that Skylar said she made for Ben a few years ago. They're having fun making roads in the sand and running wooden vehicles around them. It seems like such a simple childish game for an eight and seven-year-old, but I guess that's the draw for them. I know Matty hasn't had a lot of simple childish fun in his life and Ben's just thrilled to have a friend to do anything with.

Skylar leads me to a door at the other end of the cavern past the animal pens. Even after the tour she gave us, I'm still amazed at the set up here. The sprinklers and UV lights on timers keep her multi-tiered garden area healthy and producing. Belle will go crazy when she sees it. She's always taken care of the plants we depend on to supplement Lance's hunting, but it's always been a real challenge for her with one small grow light and lack of space. Still, she managed to keep all seven of us fairly healthy compared to a lot of other survivors, even if it meant eating fast growing sprouts at every meal.

Skylar holds the door to the shooting range open for me to hobble through on my crutch. I'm trying not to use my leg at all so it will be at its best when I leave tomorrow. I had taken a quick look into this room when she gave us the tour but now I stand and stare in complete awe as she unlocks and pulls open a double door cabinet set against one wall. There's enough weaponry hanging in racks for a small army to go to war.

She removes a handgun and a box of ammunition for it and then pulls out a shotgun and another box of shells. With sure hands she has both weapons loaded and ready for me in seconds. When she turns to me with a pair of ear protectors, she gives a little smirk and I snap my mouth shut. Yup, it's official, this is my dream girl.

She takes me through gun safety and loading procedures and I'm very thankful that she doesn't flat out laugh the first time I fire the shotgun and it almost knocks me on my butt. I have no choice but to use both my legs to steady myself while I shoot target after target until my knee is throbbing again. I'm no crack shot, but I'm sure I'll be able to defend myself if they're close enough.

Skylar goes through all the targets I've managed to hit with a slight frown.

“You’re pulling to the left with your shots but I think it’s because of your leg. Just try and keep that in mind if you end up having to shoot at anyone.”

I’m slightly embarrassed that I didn’t do better so I ask her, “Do you know how to shoot all these guns?”

As soon as I ask, a small smile flits across her face and I’m already regretting asking. She walks down the lane and puts up a new target in the silhouette of a person before coming back and reloading the handgun I had been using. In one smooth motion, she lifts the gun, squares off, aims and fires continuously until her clip is empty. She safeties the gun, sets it down and goes to retrieve the target which is further back than the ones I’ve been practicing on. With a sweet smile on her face, she hands me the target without even looking at it. I hold it up to the light and grin even as my shoulders slump. There’s a perfectly shaped heart made from holes in the middle of the silhouette’s chest and one hole dead center of the head. I let the target fall and look at her in admiration.

“Sooo, maybe I should stay here with the boys and you should go save my family.”

She shrugs her shoulders and just says, “Done.”

She would, she would go and risk her life for my family if I let her. I reach up and tuck a stray gold curl behind her ear.

“Thank you, Skylar. I really appreciate that and everything else you’ve done but I have to go.”

Chapter Twenty Seven-Skylar

Dinner's another success with the boys gobbling up the easy chicken stir-fry over rice that I made. I'm antsy and nervous about Rex leaving tomorrow and can't let myself settle down to the movie they've chosen to watch. This one's about a school for magic where a boy with a scar finds trouble with his friends. I've seen it before as well as the many that come after it so I just putter around the kitchen and think about what I'll do if Rex doesn't come back.

It's only been a day and a half since I first saw him but the feelings I have for him are real and they're strong. I don't want to live like this anymore. I want to make plans and have people around and build a life, but I don't want to do it without him. My mind's made up. If he doesn't come back, I'll lock the boys into the living quarters with a pile of sandwiches and video games and go after him. AIRIA can keep them out of trouble for a day or two if necessary.

Finally, satisfied with my plan, I go over and sit on the couch beside Rex. He gives me an absent smile before turning back to the movie but jumps slightly in his seat when I lean against his arm. This time his green eyes are all mine when he looks back to me. Those eyes are soft and sweet when he lifts his arm and puts it around my shoulders so I can lean against him more comfortably. Wow, this feels so right, like I've been missing it for all my life. I realise that other than Ben, I haven't been hugged or touched since my Dad died, and even then he wasn't very affectionate. Mom, Mom was the hugger and person who gave little touches and pats of love freely. I really miss that. I close my eyes and let myself drift off to sleep.

AIRIA's voice jolts me awake and I feel Rex's arm tighten around me. Rubbing sleep from my eyes, I peer at the clock and see that we've slept the night through on the couch but it's still very early. The boys are struggling out of their nest of blankets under their coffee table fort. I clear my throat and push away from Rex.

"AIRIA, what did you say?"

"Skylar Ross, movement detected within the perimeter."

Rex and I shoot to our feet simultaneously. We look at each other in alarm. Is it those men? Are those men back to make sure the boys died?

"AIRIA, how many?"

"Skylar Ross, two life forms detected."

Rex rubs the sleep from his face and shoots a quick look at his little brother who is trembling in fear.

"It's ok, Matty. They can't get in here. No one even knows we're here, right Sky?"

I nod reassuringly to the boy. "He's right, Matty. No one can get in here. You're totally safe! Listen, why don't you boys get this place cleaned up a bit, brush your teeth and Rex and I will work on getting breakfast ready?"

The boys calm down a bit with something to do and as they fold up their blankets and take down the fort Rex and I move into the kitchen where we speak in lowered voices.

"Do you think it's them? Did they come back to make sure you and Matty didn't make it?" I ask him.

He shakes his head. "I don't know. Why would they and what about the radiation? Why would they risk it? Is there any way we can find out? Do you have cameras or something we can see out there with?"

I frown in frustration. "No, the cold temperatures kept freezing the cameras so Dad stopped replacing them a long time ago. AIRIA, how's the weather?"

"Skylar Ross, radiation cloud front has mainly passed from the region. Radiation levels have dropped but are still elevated and may cause future health issues. Recommend covering all exposed skin to limit exposure."

Rex looks over at the boys as they finish cleaning up and head to the bathroom to wash up. He waits for the door to close behind them before speaking.

"I need to go. I need to get down to the hotel and find out what's happening with my people." He swears under his breath. “Who’s out there? I can’t just go out and shoot those guys, can I?”

I grab his hand and give him a steady look. “You might have to, Rex. Those are the bad guys. They tried to kill you! I… wait.” “AIRIA, is there any way to determine who is in our perimeter? Do you still have audio working or did that fry with the cameras?”

“Skylar Ross, body mass scans indicate a male and female. Scanning remaining exterior audio now…Scan complete, patching audio through now.”

Rex’s face is filled with confusion at the news that there’s a man and a woman outside. We both strain to hear anything that will give us a clue as to who they are but the sound from the speakers is filled with crackles and static. I’m about to give up and tell AIRIA to shut it off when we both hear Rex’s name faintly shouted from a distance. When Matty’s name is yelled next his face splits into a grin and he barks out a laugh.

“That’s Marsh! He’s looking for us! Sky, I have to let them know we’re here. I need to go out there!”

I try and smile. I’m happy his friend’s out there and not those men, but that means all this is over. My life here alone with Ben and the time I’ve had Rex to myself is over. It’s what I wanted but I’m scared and nervous for what comes next. Everything’s about to change and I’m not sure I’m ready, but Rex is looking at me expectantly so I force a full smile on my face.

“I got this, hold on. AIRIA, are they in sight of the door?”

“Skylar Ross, intruders are traveling towards living quarter’s exterior door and will be in sight in four point six minutes at current travel speed.”

“AIRIA, once they’re in sight, open exterior door and announce the location of Rex and Matty. Tell them to enter the airlock and decontaminate.”

Rex is practically bouncing he’s so excited. He grimaces as he puts too much pressure on his leg but his voice sounds thrilled.

"It's either Sasha or Belle with him! You're going to love my family, Sky. They're such good people. I should have known they'd come looking for us!"

I just smile and nod. I really am happy for him. If some of his family are here looking for him then that means Rex won't have to put himself in danger going to rescue them. The boys come bounding out of the bathroom demanding breakfast but after Rex explains what's happening they bounce around whooping it up. I don't think Ben really knows what's happening but he's really excited about all these new changes.

"Skylar Ross, airlock door opened and the announcement made. Subjects entering airlock."

Rex moves over to stand in front of the door and pulls me to his side. His grip on my hand tightens the longer we wait. The boys are practically vibrating as they wait just behind us. I'm just about to ask AIRIA what the hold-up is when the small red light above the door flashes to green before it slides to the side.

Standing in the small room is a boy around Rex's age with long blond dreadlocks tied off to the side with a piece of leather string. His eyes go big and dance with amusement and glee when he sees Rex standing with me. Beside him is a very pretty girl a few years younger than me with gorgeous red hair. Her eyes lock onto Rex and are full of relief and love before sliding across to me and then down to mine and Rex's joined hands. Her expression blanks out but her eyes look hurt for a split second before they harden.

After all the teen dramas I've watched over the years, I can see what's coming in a heartbeat. Without ever speaking to her I've already made an enemy.

Chapter Twenty Eight-Rex

My face hurts I'm smiling so big when Marsh steps into the living room and looks around with his happy, goofy face.

"DUDE, little Dude! Oh wow, pretty Dudette and another little Dude! Man, it's so good to see you! We totally thought you got nuked!" He went to say more when his eyes zeroed in on the TV and the game on the screen that the boys had been about to start playing while they waited. His face changed to disbelief and his voice came out in reverent awe.

"DUDE, is that Rally Racer 6000?"

Laughter just explodes out of me. I'm so relieved that they're here and safe. I reach out and punch him in the arm.

"Marsh, focus man! Priorities!"

He shakes his head making his dreads bounce and looks back at me with his Marsh grin.

"How did you guys find us? Is everything ok with the others?"

Marsh's face goes serious but he snags Matty's arm and pulls him into a playful headlock while he talks.

"Man, it's a total cluster! Ted's gone power mad. He snagged Belle and locked her and Dads up but Sasha and I hid out in an empty wing at the hotel. Pops hasn't been back from the scout of the ski resort but I'm guessing he's under cover somewhere because of the temp drop. I did some sneaking around yesterday to see if I could find Belle and Dads but they're locked down hard. I did find Jones. He's the one who told me what happened to you and little dude."

I look over at Sasha but she's busy looking around the room so I ask Marsh, "Who's Jones?"

Marsh lets Matty go and throws his hands up.

"He's the Wag that took you and Matty! Karma worked him and his buddy Bo over good. He's a melted man. Too many rads fried him up good. He was feeling guilty now that he's about to meet the maker so he told me how they snatched you guys and where they left you. As soon as the temp started going up Sasha and I bailed out and came looking for you two.

So… who's the *la-dy*?" He asked with his eyebrows moving up and down comically.

I was relieved that Sasha and Marsh had escaped. It would mean the three of us could work together to go and try to free Belle and Ethan. Even better, Sasha could stay here with the boys and Skylar could go down with Marsh and I. She'd be a real asset with her sharp shooting.

I tugged Skylar closer to me and made introductions.

"This is Skylar and her brother Ben. This is their place. She rescued Matty and me and brought us in here right before the radiation cloud hit. She saved our lives. Skylar, Ben, this is Marshal, but we call him Marsh for short. He's my best friend and this is Sasha. She and her mom, Belle took me and Matty in on the first day. She's like a sister to me."

Marsh gives Skylar and Ben fist bumps in greeting but Sasha only nods with a tight insincere smile before turning to me.

"Great, I'm glad you and Matty made it to safety Rex and I'm **super** happy that you made some new friends but let's go! We have to get back and rescue Mom and Ethan!"

Before I could say anything, Marsh bumps his shoulder into Sasha's.

"Yo, take a breath Sash! I told you they're fine for now. Ted's not going to hurt them. He wants Dads for his doctoring skills and Belle's locked in with him in the clinic. Dads won't let anything happen to your Mom. We need to get our game faces on and work out a plan before we launch. Besides, we haven't eaten in two days, I'm sure Skylar wouldn't mind sharing any vittles she might have handy!"

I could easily see how frustrated Sasha was by the anger that flashed through her eyes and the red spots on her fair cheeks but she again flashed Sky that tight insincere smile and spoke through gritted teeth.

"Thank you for rescuing and taking care of *my* boys Skylar. Would you happen to have any extra food you'd be willing to share?"

Food, yes that's a great idea. We'll eat and Sky and I can tell them about the plans we have to move everyone up here and then plan our rescue. I give her hand another squeeze.

"Sky, do you mind if Marsh and Sasha stay and eat something before we go back to town? We can fill them in on all the things we've talked about."

Skylar tugs her hand out of mine and gives me an almost identical tight smile to Sasha's before she nods and heads over to the kitchen. Uh, what just happened there?

Chapter Twenty Nine-Skylar

That's just frackin' great! I've got a "mean girl" in my house. I guess she missed the memo that there's no high school anymore and the apocalypse wiped away all that petty teen crap. Judging by the way her eyes and mouth tightened when Rex made that "like a sister" introduction, she doesn't feel the same way. I ignore the "Oh goody" sarcastic mumble from her when Rex says we'll tell them about our plans and head over to the kitchen to get breakfast started. I'm going to kill this girl with kindness until she lets it go. After all, she's been with Rex and Matty for the last seven years. I kind of understand her feeling territorial and I know how important these people are to Rex, they're his family.

I start pulling out everything I need for a full on fry up breakfast and see Sasha watching me out of the corner of my eye as Rex fills them in on what's happened to him and Matty over the last couple days. Her eyes almost bug out of her head when I set a bowl full of fresh eggs on the counter so I try and extend an olive branch.

"Sasha, would you like to help me make breakfast?"

I send a smile over to Rex when she turns her back to me as if she didn't hear me. He's got this clueless confused look on his face that makes me want to laugh a bit. I'm definitely not an expert when it comes to relationships with other kids my age but I have the wide wisdom of teen TV shows to draw from. I know that not everything I've seen on TV is real but I'm sure a lot of it was drawn from real life.

As soon as the bacon starts to send out its distinctive aroma, Marsh thrusts a hand up in front of Rex's face cutting him off mid-sentence and strides over to me. He leans over the counter and gives me the biggest puppy dog eyes I've ever seen.

"Bacon Goddess of my dreams, please say you'll marry me? I'll just tie Rex back up to the nearest tree and we can live here together in bacon love forever.
I…WILL…BE…YOUR…SLAVE!"

I let out a genuine laugh. This guy is seriously cute in a goofy, squeeze him and put him in your pocket kind of way. Rex elbows him aside with a fake snarl.

"Too late, finders keepers! She and her bacon glory are all mine!"

I shake my head in an amused way and send an eye roll at Sasha to try and include her. The flat hostile look I receive back tells me I'm not gaining any ground there.

The boys all descend on the video game controllers while I finish cooking the meal and Sasha walks around the room peeking through doors. I keep one eye on her as I cook and try and think of a way to connect with her. This girl will be living in my space soon enough and I don't want there to be animosity between us. She doesn't need to be my BFF but I refuse to have to tip-toe around her in my own home.

There's a minor male stampede when I call everyone to eat and the bacon induced moans from the boys fill the air above the table. At Rex's prompting, I fill them in on Ben and my story and explain some of the features of our home. Rex takes over with our idea about moving the town's groups up here to live. Marsh seems excited but Sasha just sits in silence picking at her food until she finally throws down her fork.

"Yay, super Skylar will save us all! Now can we go rescue my Mom?" pours out of her like sarcastic syrup.

Rex frowns at her tone but gives her an understanding nod.

"Marsh man, are they in danger?"

Marsh wipes his mouth with his sleeve before answering.

"I don't think so. I checked on them a few times when the guard was called away. They've got a long chain on Dads ankle attached to the wall so he can still take care of patients and Belle's free to move around the room to help out as his nurse. They stripped our rooms of all our supplies so they got what they wanted. I don't think anything will happen until the scouts get back from the ski resort. Who knows how many of those guys got caught out in the weather that locked us down. I'm sure Pops is fine. He's way too sharp for one of those idjits to get the drop on him." He looks over at Sasha and pats her hand before continuing.

"We have time. There's no point in us storming the gates without a solid plan. We should try and intercept Pops on his way back to the hotel so we're all on the same page and he's not ambushed. We also need to talk about how we go forward once we get Belle and Dads back. Skylar, it's righteous of you to offer a spot for us here but it'd be pretty tight quarters with all of us living here."

Rex and I share a knowing look before I push back from the table and stand.

"I don't think space will be a problem. Why don't you let Rex and I show you around a bit? There's also some things here that I can give you to help with the rescue."

Ben and Matty asked if they could stay and play more video games so I said yes but warned Ben that his chores still needed to be done after the tour. Rex and I led Marsh and Sasha into the cavern and showed them the grow area and animal pens. While we were there I collected all the eggs from the hen house and left them by the living quarter's door in a basket. The cow would have to wait a bit for her milking.

Marsh was pretty excited about everything he saw but Sasha just scowled and followed behind us with her arms crossed. I decided to just leave her to her misery for now. Until her Mom was safe she wasn't likely to change her attitude. I palmed open the door to the connecting tunnel and listened to Rex and Marsh talk about how cool it would be to skateboard in the long stretch of corridor. The barracks had the same effect on Marsh as it did on Rex the first time he saw it. He walked around in amazement as we showed him all the supply rooms and huge kitchen and washrooms. When we opened the door to the massive grow room with its empty beds and tiered shelving filled with pots waiting for plants he whooped in excitement.

"Sasha, your Moms will go spaz for all this space to grow her plants!" He spun around. "Sash? Sasha? Hey, where'd she go?"

We backtracked through the room looking into doorways, but we couldn't find her. I was silently fuming at her pouty behavior as we made our way back through the tunnel. Rex was starting to limp more even with his crutch so we'd have to wrap it up and get him on some pain killers before we set out for town. I had almost thought I could go with them to help if Sasha stayed back with the boys but she clearly wasn't going to go for that.

We made it back to the cavern and after calling her name and getting no answer went to the door to the living quarters. I couldn't believe how immature this girl was being. I stopped at the door to grab my basket of eggs but they were gone. A smile crossed my face. Even with a buddy to hang with, Ben was keeping up with his chores. Except he wasn't.

The boys were exactly where we had left them, playing video games on the couch, and Sasha was nowhere to be found.

Chapter Thirty-Rex

"She left," Matty replied when I ask him if he's seen Sasha, never taking his eyes from the screen. I walk over and hit the power button on the TV to get his full attention.

"Matty, where is Sasha?"

He looks sideways at Ben before shrugging his shoulders. "I don't know. She came back not long after you left and said she was going to get Ethan and Belle. She said you guys were going to meet her there." He looks past me at Skylar. "That was really nice to give Sasha all those eggs for the people at the hotel. You're really nice Skylar." The kid blushes bright red and looks down at his feet.

I spin around and see Sky's face has gone chalk white and her eyes are cold with fury.

"Sky, what is it?" I know Sasha's desperate to get her Mom back but I don't know why Sky's so mad about her leaving.

"SHE TOOK THE EGGS." Each word of that sentence could cut glass they're so sharp. Now I'm even more confused. Sky's mad over a few eggs? She's been so generous to us that I don't see her being this mad about a few eggs. When she sees my confusion she explains.

"Your "like a sister" is going to sell me out! She's going to go and bargain with that Ted guy for her Mom. The eggs are the proof! She'll tell him what we have and where we are in exchange for her MOM!"

My mouth falls open in disbelief. I know Sky has trust issues but there's no way Sasha would do something like that. She might try and barter with those eggs for Belle's freedom but she wouldn't sell Sky and her brother out. Marsh is just standing and shaking his head. So I try and reassure Sky but before I can say a word she's already talking.

"AIRIA, when did Sasha leave? How far has she gotten and why the HELL did you let her OUT?" She's furious and not looking at me at all as she starts pulling outerwear from a closet. When she pulls her rifle from the closet my mind goes blank.

"Skylar Ross, Sasha Bennett left by the living quarter's exterior door thirty-four minutes ago. She crossed the southern perimeter sensor boundary eighteen minutes ago. Access level red restricts interior movement and entrance. It does not restrict egress, with the exception of Benjamin Ross, as specified by you."

Skylar groaned in frustration but went back to loading her rifle. I don't know what to do. Sasha wouldn't do what Sky claims but I don't know how to make her believe that and what the hell does she plan on doing with that rifle?

"Sky, Skylar! What're you doing?"

The look she gives me is cold and it's like she's looking at a stranger.

"I'm going to stop her."

Marsh jumps between us with his hands up.

"Whoa, whoa, everybody take a chill pill! I'll go and get her. She's got less than an hour's head start so if I motor I can reach her before she hits town. Rex, Skylar, get ready. Get that knee wrapped up and any weapons you're willing to lend us so as soon as I get her back we can go. Skylar, I promise you I'll stop her even if it means tying her to a chair here while the three of us goes and gets the rest of our people. But, I honestly don't think that's what she's going to do. I think she's just crazy with worry for her Mom."

Skylar just stares at him for a moment and I think she's going to push past him before she finally nods and goes to hand the rifle to Marsh but he backs away shaking his head.

"Uh, thanks but I'll need more than a quick lesson before I use one of those. Besides, I'm just bringing her back not shooting her."

I see Skylar roll her eyes as if she wants Marsh to shoot Sasha. Have I been wrong about this girl? Are her trust issues too big for us to overcome? The fact that she immediately thought the worst of Sasha even after I vouched for her says maybe they are.

It takes Marsh minutes to gear up and head out. I stand in the middle of the living room facing Skylar in the silence. I don't know what to say and she's not talking. Based on her cold expression there might not be anything left to say.

Interlude

Ted

“Jones is dead, boss.”

Ted Redford grunts in annoyance as he stuffs more supplies into a backpack.

“Good, that loser and Bo couldn’t even do the simplest task. Any luck finding the girl or that other punk?”

Mickey’s face is hard when he shakes his head.

“No, the men searched the whole hotel but came up empty. Those kids must’ve ducked out right away.”

Ted tosses the bulging bag at Mickey.

“Well, doesn’t matter now, although it would have been nice to have a second woman with us at the lodge. Let’s get a move on. I want to be clear of town before either the generator dies for good or anyone who survived the scout run makes it back here. Either one of those things can happen at any minute and I don’t plan on getting caught here with these losers.

It’s a good thing it was you that found that hunting lodge and not any of the men or we’d have to take them with us and dispose of them along the way.” He looks around the stripped-down office and shakes his head. “I’m sick of sharing and carrying the dead weight around here. Grab the woman and I’ll meet you out back by the trailer.”

Mickey grunts again and shoulders the pack Ted had thrown at him before leaving to collect Belle.

Ted takes one more look around and smirks before leaving. It was a good gig while it lasted but he wasn’t going to go down with a sinking ship. All the gas from every vehicle and every tank had been drained of its last drop. When the generator failed there would be no way to pump the water that kept them going here for so long. Add to that the very good chance that those savages would be headed here to attack and he was cutting his losses. Now that he had that pansy soldier boy's supplies and woman as well as a new secure, hidden base, he was laughing.

Sasha

Sasha stumbled on the uneven frozen ground in her haste and cried out when one of the eggs tumbled out of the basket and smashed open in front of her. She pushed to her feet and dashed a coat sleeve across her wet eyes. She was furious, frustrated and heartbroken all at once. She couldn't believe Rex and Marsh. After all these years they were more interested in a stupid girl with her stupid bunker than saving her mom. Didn't they care that Ted could be hurting her right now? She would show them. She would take the eggs and the location of Skylar's cave of wonders and use them to bargain for her mom's freedom. Once Ted knew what that girl had hidden away he'd let her mom go and ***he*** could move everyone up there. He and his men would be so busy moving everyone up the mountain that he wouldn't give them a second thought. Rex, Matty, and Marsh would be on the way down to look for Lance so they wouldn't be in Ted's sights when he got up to the girl's door.

If Skylar wanted to share everything she had with everyone and be the big hero she could start by dealing with Ted and his men.

Sasha finally caught a glimpse of pavement and breathed a sigh of relief as she passed the last of the dead trees and stepped down onto the road. She finally knew where she was. She turned towards town and took two steps before stopping short. Just ahead of her and coming in her direction was Ted. He was pulling her mom along by a rope tied around her bound wrists. Just behind them was that big lug Mickey and he was pulling a small trailer that was heaped with goods covered by a tarp. She took a deep breath, squared her shoulders and started walking towards them. She didn't know why they were out here but it would save her a long walk back to town.

Marsh

His heart bashed against his ribs and his breathing was a ragged rasp. He'd been throwing himself full-tilt down the mountain to catch up with Sasha. He'd like to believe that she wouldn't do what Skylar accused her of, but he'd seen the ugly jealous look in her eyes every time she looked at Skylar.

Sasha was only eight when the bombs dropped and she'd grown up listening to Dads' and Pops' constant reminders to not trust anyone else and always put their group first. It wasn't impossible that one jealous fifteen-year-old would do exactly what Skylar had said.

The road was just ahead, he could see it through the trees. Once he hit pavement he'd make even better time and should spot Sasha up ahead. He reached out and grabbed the last tree before the pavement started and used it to pivot towards town. The glimpse he saw of the road had him swinging all the way around and back in the direction he had just come before diving down behind some dead bushes.

He lays there and tries to catch his breath and figure out what to do. He just barely caught a look at the people in the road but Belle and Sasha were easily recognised by their red hair. So was Mickey from his size and sheer ugliness. Marsh could only assume the other man was Ted. That was confirmed a few minutes later when he heard Ted bark at Sasha to move faster as they passed close by to his hiding place on the way back up the mountain.

What should he do? He could follow Sasha and Ted back up to Skylar's and try to beat them there or he can try and free Belle from Mickey. He really wished he had taken that rifle from Skylar now. The only other thing he could think to do was get back to town and try and get help from his Dads and Pops if he'd made it back.

Marsh peeps around the bushes and sees Mickey sitting on the trailer with Belle at his feet on the pavement. Ok, they're not going anywhere until Ted gets back which will take at least an hour and a half. That is if Skylar doesn't shoot him as soon as the door opens so town it is. He needs help and he needs to let his Dads know what's happening.

Marsh carefully moves deeper into the dead forest until he's far enough in not to be seen from the road before paralleling it to get around Mickey and Belle. As soon as he's far enough along he leaves the forest for the road and then runs as fast as he can back to town.

Lance

"We have to move faster!" Lance yells at the ragged group of refugees that he had convinced to flee the hotel.

As soon as the temperature had started to rise this morning he had left the ski chair operator outpost building that had saved his life. The scouting trip had gone wrong after the second day when one of Ted's men had tried to put a knife in his back. That guy had ended up with an arrow jammed through his own neck for his effort. Lance had already seen enough of the group of savages that were living in the resort to know they were a solid threat. The screams of their victims echoed across the empty ski runs as they were cooked over huge bonfires and then consumed. Lance had to get his family away from the tempting target the hotel was to this group.

He should have been back three days ago but he was forced to take cover when the temperature started to drop so quickly. He knew exactly what it meant from previous years when a radiation cloud had passed through. It was sheer luck that he learned of an even worse fate. He had to pass fairly close to the resort to make it back to town and saw the group of savages massing for an attack. He got all the information he needed on their target when he came across one of the animals alone in the trees relieving himself. After getting what he needed from him he broke his neck and left him to the animals before sprinting for home.

What he found at the hotel was chaos. The generators were dead, Ted and Mickey were gone and his husband was chained by the ankle to a wall. No one had seen any of the kids for days and Lance didn't know where to begin to look for them. He had to push the devastation he felt to one side and deal with what he could control. He came very close to just taking Ethan and running, leaving everyone in the hotel to their fate but his husband convinced him that there were still good people here, not to mention children.

He yelled and screamed and bullied them to get them all moving and out of the hotel and town. They headed north just because it was away from where the savages were coming from. Lance and Ethan figured they would get everyone far enough away and then regroup and try and find some kind of shelter that would temporarily hold everyone.

Lance had to look down at his feet in order to not yell at them to move faster again. He was starting to think that he and Ethan would have to leave the group behind for their own safety when a murmur flowed through the crowd. His bow was off his shoulder with an arrow notched in an instant as he scanned in all directions for the threat he knew was coming. What he saw instead was his son sprinting towards him.

...end interlude

Chapter Thirty One-Skylar

Marsh had been gone for over an hour and I hadn't spoken to Rex once. Realistically, I knew Sasha's possible betrayal wasn't his fault but I had finally let myself trust someone and it was hard to swallow. Besides, he seemed to be just as mad at me from the way he reacted. As if I wasn't supposed to worry about my home and brother falling to men who had no problem leaving children to die in the woods.

There was no way I could just sit and wait to see if Marsh would bring her back so I kept busy cleaning up breakfast dishes, giving the boys quick showers and milking Nods. I swear I must have wiped the same area of the counter in the kitchen ten times before AIRIA finally announced two people, male and female had entered the perimeter.

Rex jumped to his feet and pumped a fist in the air.

"I told you he'd bring her back! Now you'll see she wasn't going to do what you said."

I just shook my head at him. "AIRIA open the airlock and let them in. There's no need to decontaminate."

I was in too much of a hurry to hear Sasha's explanation to worry about such a small amount of exposure. They'd only been out there for less than two hours and the cloud was gone.

Rex and I watched the red light above the door in silence. The boys had a movie playing at a low volume but their eyes were on us as we waited. Finally, the light turned green and the door slid to the side. Sasha came flying through like she had been pushed from behind and I jumped back so she wouldn't crash into me. My eyes shot back to the door to see why Marsh had been so rough with her and instead found a gun pointed at my face and Marsh wasn't holding it.

Rex yelled out and backed away when he saw who was standing in the door. I didn't even need him to tell me that this was Ted. Rex looked from him to Sasha who was getting to her feet with total confusion on his face.

"Sasha, what did you do?" He asked chokingly.

Sasha got to her feet and sent Rex a look filled with apology.

"I did what I had to, to save my mom. She's all I have." She turned to Ted and stood up taller. "There, I kept my end up, now can I have my mom back? You promised!"

Ted let out one of the ugliest laughs I've ever heard. There was no surprise at all in me at what he told her.

"Now why would I do that? I now have the keys to the kingdom and some pretty little princesses to keep me company. No, I think I'll just let Mickey keep your Mom as a consolation prize for me dumping him." He laughed again as Sasha dropped to her knees and started to weep, then turned to me. "You must be Skylar. Just so you know - Sasha doesn't like you very much. Nice digs you got here. Greedy little girl to have kept all this to yourself but Sasha tells me you decided to share after all. That won't be happening, at least not with anyone but ME!"

I'm numb, I feel nothing. This is exactly what Dad had said would happen if we tried to help others. My voice is flat when I speak.

"AIRIA, voiceprint Ted. He is an intruder and threat to our safety. Lockdown."

Ted's looking at me with contempt but starts swiveling his head around as thunking noises come from every door in the living quarters. There is now no way he can enter the cavern or exit to the outside without my permission.

"What just happened? What did you do?" he yells at me.

I give him a level look, free of fear.

"Did you really think you could just walk into my home with your little gun and take over? Did you really think it would be that easy? This is a military grade bunker, run and maintained by an artificial intelligence system. Do you think you can negotiate with a computer? We are now on lockdown with all doors locked and sealed. You will have no access to any of the supplies or technology in this bunker." I half turn away from him and address the ceiling. "AIRIA, monitor vital signs, any harm to myself or Ben…" I catch Rex's agonized look but look away, "Or Matty, and you are instructed to flood this room with a lethal amount of gas to terminate the threat."

I turn back to Ted who's not looking so cocky anymore but still points his gun at me.

"We go, you go."

He and the others have no way of knowing that I'm bluffing, at least I think I am. As far as I know AIRIA doesn't have that capability, but I'm not a hundred percent certain. The fact that she didn't acknowledge my command makes me think I'm right.

Ted waves his gun around at the others before grinning.

"Well, it looks like we're in a stalemate then. I can't hurt you and the kids without dying, but I'm not leaving with nothing, so let's just get down to negotiations then. What's stopping me from shooting your friends here to get you to co-operate? You didn't mention them to your fancy computer."

I look down at Sasha with disgust and pull her to her feet.

"Go for it. Do what you want with this one. She had no problem putting two children in danger to get what she wanted. As for Rex, well, I'll feel really bad about it but he brought this viper into my home so his fate is not my problem."

I don't let myself look in his direction when I say such hateful words. I can only hope he knows I'm bluffing and trying to stall for time. I just need for Ted to relax and let his guard down. If I can get to my rifle or get the boys out of this room I might be able to do this monster some damage.

"You know, I think I believe you. You're pretty cold, I like that in a woman. So, where do we go from here?" He asks before giving Sasha a shove towards the back of the room and out of his way.

I don't answer him as he starts to poke around the room. He heads over to the kitchen but keeps his gun pointed in our general direction. When he looks away for a second to look into a cupboard he opens, I wave the boys closer and take a step towards the back wall. I'm frozen in place when he glances back.

"Got any grub in this place?"

"Sure, there's plenty in the fridge and even more in the pantry beside it," I tell him.

He gives me and the others a menacing look and waves his gun at us again before turning and opening the pantry door. While he's looking inside, I nod at Rex towards the back wall where the door to the cavern is and he immediately moves closer as I take another step back. When I see Ted turning towards us I call over to him.

"There's cooked bacon in the fridge left over from breakfast."

Ted lets out a whistle of admiration and immediately turns and pulls the door to the fridge open and bends over to look inside.

This is it, this is the time.

"AIRIA, all lights off!" I say just above a whisper followed by "AIRIA, open living quarter cavern entrance door!"

The lights go out and I hear the thunk of the lock opening before the sound of the door sliding open. I'm spinning and rushing to where I think the door is. I barrel into one of the boys and push him ahead of me before smacking my elbow hard on the side of the door. The boys and Sasha are crying out and Ted's yelling behind us. I feel strong hands shove me through and I fall to the floor of the cavern just as there's a flash as Ted's gun goes off. I catch a fleeting glimpse of Rex's outline in the door before he drops. I think he's made it through so I yell.

"AIRIA shut the door!"

As soon as I hear it slide close I yell for all lights on. It takes my eyes a moment to adjust again and when they do I let out a scream of defeat. Sasha and Matty are cowering on the floor beside me but Rex and Ben aren't here. What causes my heart to almost stop is the fine spray of red droplets that are on the floor between me and the door.

I gotta think, because that's blood. I've got to do something…BEN!

I scramble to the door on my knees and rest my palms and forehead on it. What now, what do I do now?

"AIRIA, how are the vital signs for Rex and Ben?"

"Skylar Ross, Rex Larson's vital signs indicate that he is in distress. Benjamin Ross has an increased heart rate but all other signs are within normal range."

Ok, ok Ben's fine so that's Rex's blood but how bad is it and what can I do?

"AIRIA, can you tell how bad Rex is hurt?"

"Skylar Ross, Rex Larson's vital signs indicate he is in distress but are not life threatening at the moment."

I take a deep breath. Ok, he's ok for now but what will Ted do?

"AIRIA, patch me through to the living quarter's speakers."

"Skylar Ross, channel open."

I can hear them. I can hear Ted ranting and swearing and glass breaking. Underneath that, I can hear Ben's muffled crying and what I think is Rex moaning.

"Ted, TED! Can you hear me? Shut up and listen to me!" When I can't hear him anymore I continue. "Same deal applies except now it includes Rex. If you hurt either of them or Rex dies from his wounds, I'll gas you! So you better give him a towel to stop the bleeding. Now, I'm going to put you together one hell of a care package and you're going to take it and leave."

I wait for a reply and it comes quickly. "Why the hell should I believe you after that stunt you just pulled? You list off everything I'm getting and I'll let you know if it's good enough!"

I wrack my brain for the magic item that's going to make him happy. Food, water, guns, comfort items all that's easy but what will make it irresistible to him? I'm trying to think of something, anything when a hand clutches at my arm making me flinch. It's Sasha, her face is smeared with snot and tears.

"Just give him everything. Get Rex out of there and let him have it all!"

I growl at her and shove her back to the floor. I practically spit at her.

"Shut up! You shut up. This is all your fault, all of it! You get no say in anything! That's my brother in there and this is MY home so you SHUT THE FRACK UP!!!"

She pushes herself back away from me with a white face and gathers Matty up against her like a shield so I turn back to the door. I open my mouth to speak and then close it. Frack! The speaker was still on so that means Ted heard all that. Well, he's dead wrong if he thinks I'm giving him everything. When this is all done they're all out, all of them! It's been just three days since I let outsiders into my home and my trust and look what happened. No, it's going to go back to the way it was before and I'm going to forget there is an outside.

"Ted, are you listening? Here's your list. Food, water, guns, blankets, booze and survival gear. That's what I'm offering."

I hear Ted bellow out a laugh. "Are you nuts? You think I'm going to take the few bags of goodies that I can carry and leave all this?"

My mouth hardens and I lean my head against the door.

"No Ted, you're going to take as much as a pickup truck can carry and drive it out of here."

There's silence coming through the speakers for what seems like a thousand heartbeats before he finally responds.

"DONE! But Skylar, the boy stays with me until I'm in that truck!"

I let out the breath I was holding and sag against the door.

"I'm going to need time to fill the truck so make sure Rex's bleeding is under control. He dies, you die."

I hear him say, "Yeah, yeah, yeah" through the speakers so I have AIRIA disconnect. I pull myself to my feet and slowly turn around. With the deal made all the rage drains out of me and I'm left tired and feeling beaten down. I look down on Sasha dispassionately but frown at Matty's scared tear covered face.

"Matty, I'm going to get Rex out of there. He'll be fine."

He looks up at me with big sad green eyes that perfectly match his brothers and asks, “And Ben too? Ben’s going to be ok too, right?”

I give him a reassuring nod but my voice is weary.

“You bet, buddy. Rex and Ben will be just fine.”

Sasha watches me warily, waiting to see what I do next. I can’t wait to never see her face again, but first, there’s work to do.

“Get up. Get up and help me load the truck so we can end this.” There’s no bite to my words but they’re firm.

She pushes Matty away and gets to her feet, a spark of anger coming into her eyes.

“Why should I help you? Do it yourself!” I just stare her down and wait. Sure enough, she’s got more to say. Her face takes on an ugly sneer. “Like you said, this is your house. Skylar Ross, perfect girl in her perfect home. Not so perfect anymore, is it? How’s it feel to have to deal with reality, cuz this is the kind of stuff the rest of us have had to deal with since the world changed. You think you’re so special? You think you can just swoop in and take my family, take Rex?”

Ahhhh, there it is. The real reason all of this has happened. The envy and jealousy of a fifteen-year-old girl. I really, really want to just punch her in the face but it’s just not worth it so instead I try and put it in perspective for her.

“Yup, that’s me living the dream! It was a perfect life while I watched my mom bleed out in front of me the first day. I was ten. It was even more perfect when one of the people from your town shot my dad in the back, for his boots. I was thirteen and all alone with a two-year-old. Yup, perfect. It was so perfect that I decided to risk my life and my brother’s and my home to help your people.

“Sasha, I never did anything to you but try and help. You came into my home and acted like a spoiled child. You were so jealous and petty over a BOY that you brought a dangerous man here and put two little kids and the rest of us in danger. For what? For revenge? To get your mom back? How’d that work out for you?” I shake my head at her in exasperation.

"You will help me get that truck loaded because I'm done with your sick, sad, pathetic drama or I'll just sweeten the deal with Ted by including YOU." At her horrified look, I give her a cold smile. "After everything you've done, I wouldn't even feel bad about it. Now let's go!"

I turn and start walking away, I would never actually turn her over to Ted but she needs a wakeup call because this IS reality and I'm living it.

The first thing I do is go to the gun safe in the shooting room and arm myself with a handgun and rifle that I sling over my back. I don't plan on going back on the deal but I won't be double-crossed either. Sasha doesn't speak to me again as we start loading supplies into my Dad's old pickup but the misery on her face the whole time makes me hope that my words have gotten through to her. The truck's been covered by a tarp for years but I always started it every month like he used to do. There are double wide doors at the back of the cavern with the same rock façade covering them on the outside as well as a smaller man door. Last time I was around back on the outside was over a year ago but I could still make out the faint track Dad used to drive in. With nothing growing it should still be manageable, just covered in snow and ash.

I don't care about most of the supplies we're loading up. There's so much in storage that this small amount won't be missed, but the guns are another matter. I hate the idea of arming this guy and any like-minded people he has. It's not right to give him the means to prey on other people, so I take the time to pull the firing pins on the five rifles I'm giving him and then empty the bullets from all the ammunition boxes and refill them with empty casings. I figure his greed will make him careless and he won't discover what I've done until he's long gone.

The truck is loaded as full as we can get it. I look around the cavern and sigh. I'll have to let Ted come through here to get to the truck and that may be too much for his greed to handle. I dig around and find the small cages Dad used to bring the chickens here back on the first day and wrangle three of the laying hens. It's the last thing I can think to do to satisfy him and get him out of my home. Once they're loaded in the back seat of the truck I pull my gun and start walking back to the living quarter's door. It's time to end this. I send Matty and Sasha into one of the storage rooms to keep them safe and out of the way before getting AIRIA to patch me through. AIRIA has been giving me updates on Rex and Ben's vital signs every ten minutes so I know they're both unharmed but I won't feel a hundred percent until I see them with my own eyes.

"TED! The truck is ready to go. The door will open in a minute. Are you ready?"

His response is immediate.

"Well, it's about time, sweetheart! Now, we ain't gonna have any more funny business! I've got my gun planted in your little brat's neck so watch yourself!"

I push away the instant terror that floods me at his words. He wants what I've got to give so hopefully he won't do anything stupid.

"AIRIA, open the living quarter's cavern door."

I'm a good twenty feet back from the door with my gun firmly in my hand. The door slides open and a second later Rex stumbles out. He's clutching a towel against the top of his left shoulder and it's stained with blood but he's on his feet and moving so I look past him as he moves out of the way. Next Ben comes with one of Ted's hands gripping his shoulder and the other pressing his gun against his little neck. Ben's eyes almost bring me to my knees. They're completely blank. My baby brother has checked out and gone somewhere else. He's been so protected his entire life that this must have been too much for his young mind to handle.

Ted pushes Ben ahead of him into the cavern but comes to a stop and pushes the gun harder against Ben's neck when he sees me with the gun I'm holding. I steel myself not to react when Ben doesn't even flinch at what must be the painful pressure of the gun.

I nod at Ted and motion him forward with my gun.

“Just to keep you honest, I thought I should level the playing field.”

He smirks at my words and starts Ben moving again. His eyes take in every inch of the cavern and he’s shaking his head in wonder. I keep backing up and we make it over the foot bridge before he stops again. His eyes are on the animal pens and they gleam with calculation. I try and beat him to it.

“I’ve put three laying hens in the truck already so as long as you don’t decide to have a chicken dinner you’ll have eggs for as long as you keep them healthy.”

He purses his lips in consideration.

“Hmm, look at all of this. It just doesn’t seem right that you’ve been keeping all this for yourself and the brat while the rest of us live in the filth outside.” His eyes leave the pens and flash back to me filled with sick amusement. “I took a look around your place while I was waiting and what a surprise to see someone I recognise. That must be your dad in all those pictures you got framed up in there.” He shakes his head with fake sadness. “It’s really too bad. If I had of know he was sitting on this sweet setup I might have let him live long enough to get us in here. Oh well, his boots and rifle came in handy for a few years until they wore out.”

Time freezes. I’m thirteen and Benny’s just a baby crying for his daddy again. I’m all alone and I don’t know what to do. I just keep stacking those rocks over his body because I’m not strong enough to dig a proper grave.

Rex’s roar of anger snaps me out of it and makes Ted swing the gun away from Ben’s neck to point it at him. My hand moves before my brain is even working. My gun’s up and level without the slightest tremble. Ted’s eyes dart to me and they have time to flare wide before I pull the trigger and a small neat hole appears in the middle of his forehead.

My brain catches up just as he falls to the rock floor and that’s when the trembling starts. I killed him, not a target shaped like a person but a real living person. No matter how horrible that man was, he was still a person and I just killed him. I sink to my knees and the gun clatters to the rock floor.

My eyes are locked on to Ben who just stands there. His eyes are still blank and he makes no move to come to me.

It's Rex who breaks the moment by scooping him up and carrying him with his one good arm the ten feet to me. He sets Ben down in between us and wraps both his arms around us and just holds on.

We stay like that with Ben's head tucked under my chin and my tears wetting his hair until Matty shouts his brother's name. Rex kisses the top of Ben's head and looks questioningly into my eyes but I've got nothing for him right now. I just lower my head down onto Ben's so he lets go of us and goes to his brother and Sasha who have come out of the storage room.

I rock Benny back and forth on the floor and whisper my love to him until I feel him move. I pull back and look down at him. A wave of relief washes through me when I see life behind his eyes and a small smile cracks my face when he says in a small voice that he's hungry. That's good, that's normal, that I can handle. I pull him up with me as I get to my feet and heft him into my arms. His skinny legs wrap around my waist before he lays his head against my shoulder. As I walk past Ted's body and Rex and the others I flash back to Ben being a baby and how I used to walk the cavern with him in a sling against me. I want to go back. I want us to go back to how it was because if this is the living I thought I wanted, I was wrong.

Rex, Matty and Sasha follow us back into the living quarters where I sit Ben down at the table and head to the kitchen. I pull out food from the fridge and start making sandwiches. I can see Rex staring at me with concern but I don't care. I've been pushed and stretched beyond anything else right now. My brother's hungry so I'm making sandwiches.

Rex steps towards me and reaches out his hand.

"Skylar…"

Before he can say more, AIRIA interrupts him.

"Skylar Ross, multiple perimeter breaches detected."

I just keep making sandwiches. Maybe I should add some cut up veggies to his plate but he doesn't like them raw so no, not today.

Rex is trying to ask AIRIA questions but she won't answer him because he has no authorization. I add a few oatmeal cookies to the plate and set it in front of Ben before going back to make more when Rex grabs me by the arm.

"Sky, snap out of it! We need to know who's out there!"

I look at him with no expression at all. I don't care who's out there but he's in my way so I ask in a dead voice.

"AIRIA, how many?"

"Skylar Ross, sensors indicate one hundred and eighteen life forms within the perimeter."

Rex sucks in a surprised breath at such a huge number before a grin tugs at his lips.

"Marsh, Marsh brought everyone up!"

I shrug my shoulders and shake off his hand on my arm before sliding past him back into the kitchen. Rex follows me and keeps on talking.

"This is great! He wouldn't have come up here unless he had Ethan or Lance with him so they're probably all here."

Sasha pipes up. "Do you think they have my mom with them? Ted left her down on the road with Mickey. They would have rescued her, right?"

Their voices are like a drill in my ear as I make more sandwiches. I just want them to stop. Stop talking, stop being in my way, just stop and leave me with Benny in peace.

"You should go see. Here, I'll make these sandwiches to go. You can take them with you."

Rex and Sasha stop talking and just stare at me. Sasha looks guilty and Rex has hurt confusion on his face.

"Sky, Skylar? Don't you want to come? You could meet the rest of our people. Help get things organized, for our plan?"

I start wrapping up the food I've made in plastic wrap and keep my eyes down.

"You should go. They're probably worried about you."

He shakes his head.

"Skylar, I'm so sorry for what happened here but Ted was a bad man! The rest aren't like that. They're good people!"

My eyes flash up at him in a brief moment of anger.

"You mean like Sasha?"

I lower my eyes just as quickly. I'm not doing this. I'm not going to argue with him. I have nothing left inside for him or anyone else except Ben.

"Really, you should go, have that shoulder checked out and cleaned up by Marsh's Dad. He's a doctor, right? I'll get you a first aid kit to take."

Rex's voice is full of anguish when he says my name and I can't take it anymore.

"GO! GET OUT! I don't want you here anymore!"

I throw the sandwiches in a plastic bag and rush around the counter to the closet as he stares in disbelief. I pull out all of their outerwear that I had washed and throw it by the exterior airlock door before I head out into the cavern to storage. I grab a pre-loaded med kit and then snag two go-bags that Dad had made up years ago. I can't remember everything that's in them but I know they're filled with survival gear.

When I carry everything back into the living area, Sasha and Matty are dressed to go and Rex is struggling to get his jacket on over the towel he has wrapped around his wound. I dump the bags at his feet.

"There's lots of survival gear in those bags and this med-kit is full so you should have everything you need."

Rex gets his jacket on and zipped up before meeting my eyes with a hard stare.

"No, that's not all that I need. You, Sky, you are what I need! I'll go but I won't leave. I'll never leave you! You're not a quitter Sky, you don't give up and neither will I. I'll be waiting for you, no matter how long it takes."

I look into those green eyes and see love and determination. He'll fight for me. But a quick glance over at Ben who just sits and stares at his uneaten food pushes all that away.

"AIRIA, open living quarter's exterior airlock doors."

One last look at what might have been and then I turn my back to him. I stand strong and rigid until I hear the door slide shut and then I let the tears flow. I go scoop up Ben and settle us both on the couch. I sit with my future cuddled against my side and let all the rest go.

To be continued…

Read on for an excerpt from Rain and Ruin, Book Two in the Endless Winter Series

http://www.theresashaver.com/books

Coming Soon….

Sun & Smoke, Book Three in the Endless Winter series.

To be notified of every new release, sign up for my New Release Newsletter at

www.theresashaver.com

You will only receive an email when a new book is coming out!

Also by Theresa Shaver

The Stranded Series

Land – A Stranded Novel
Sea - A Stranded Novel
Home - A Stranded Novel
City Escape - A Stranded Novel
Frozen - A Stranded Novel

Endless Winter Series

Snow & Ash
Rain & Ruin

Follow the link to read more:

http://www.theresashaver.com/books

Excerpt from Book Two Rain & Ruin, An Endless Winter Novel

Skylar

I make my way through and around the groups of huddled people.The smoke from so many fires that are keeping them alive in this cold is like a haze clinging to the ground. I keep my head down and don't make eye contact with anyone. I'm not ready to talk to Rex yet. It's been three days since I exiled him, Sasha and Matty.

I wouldn't even be out here if not for Benny. It only took him a day to come back to his former self from the shock of what happened. He's had nightmares every night but the rest of the time he seems fine. He started asking about Rex and Matty and his new friend the "super cool" Marsh right away and wouldn't let up with the questions on where they were, when they would be back and when he and Matty could have another sleepover. I finally had to tell him the truth to get him to stop badgering me and breaking open the scar I carry from losing Rex and the future I thought we'd have. He didn't take it very well.

At first, he just kept trying to convince me to let them come back and when I wouldn't budge he started asking AIRIA questions. When she informed him of the group sitting outside our door he got mean. That seems funny that my seven-year-old sweet brother could be mean to me but his words were like constant arrows to my heart. He's smart, that kid, he gets way more than I give him credit for.

After he said I was a coward and selfish and that I was going to let all those people die, I ran. His words were like hearing a long ago echo of the things I had said to Dad. I ran to the barracks and tried to lose myself with a hard run but his words and Rex's face wouldn't leave my mind no matter how hard I ran. So I just stopped running and faced it all.

The plan Rex and I had to move all those people into the barracks would still work but there would need to be a few changes. They could have all of it but no one would be allowed into our area. I would authorize AIRIA to provide minimum support for them in the barracks and they could have all the supplies but I was out. I wasn't going to be a part of anything they did over there. I won't take another chance on Ben's safety no matter what he says to me. Everything in the barracks was theirs except for one room, the armory. That room will stay locked down with no exceptions.

So here I am, walking through more people than I have seen in seven years and I won't lie. My skin was crawling and the hairs on the back of my neck were lifted. I get to the end of the campfires without seeing Rex or any of the others but there are a few tents and tarps set up that they might be in. My dead forest has changed so much since I was last out here. Where it use to have trees everywhere it now has stumps surrounding the area the group was camping in.

I pick the widest stump I can find and take a deep breath and square my shoulders before stepping up onto it and turn to face the crowd. No one is looking my way and I'm not going to yell for attention so I pull my handgun and fire a shot into the air. There are a few startled screams but most people are too cold to move. Every eye turns my way and all sounds cease except the hiss and crack of wood burning in the fires.

I see movement as a few people dash out of a tent and there's Rex and Marsh followed by Sasha, Matty and some older people I don't recognize. Rex's face breaks out into a huge smile when he sees me but I'm not out here for that, so I look away and focus on the air just above the crowd.

"My name is Skylar Ross. My home is behind you inside the mountain. Beside my home is a huge empty barracks. It has plenty of room for all of you with electricity, heat, hot water and all the food and supplies you need to live comfortably. It's yours. I don't care what you do in it or how you run it. There is only one condition. Your former leader, Ted came into my home and put a gun to my seven-year-old brother's neck. He threatened us and tried to take my home from me. I killed him. The same thing will happen to anyone who tries to gain entry into my home. You stay in your side and I'll stay in mine. Just in case you think you can overwhelm me, you should know that there is an A.I. computer that controls all the mechanics of the barracks and I won't hesitate for an instant to order it to cut all of it off if I feel threatened."

No one moves and no one speaks when I pause for breath. "There are doors that will open shortly just to the north of you. I suggest you gather your things and go inside them."

That's it. That's all I have to say to these people. They can figure out the rest with AIRIA's help once they're inside. I'm about to step down from the stump when the alarm from my communicator blares out followed by AIRIA's voice.

"Skylar Ross, a meteorological anomaly has been detected. Seek shelter immediately."

I look up at the sky and frown in confusion. The temperature's not dropping, if anything it actually feels a little warmer and the sky is the same boiling grey black it's always been. I don't see or feel a problem until there's a sting on my cheek. I reach my hand up to rub it away when something hits the back of my hand and it too starts to sting and then burn. I look closely at my hand but all I see is a drop of water. Then I understand, it's raining, it's raining for the first time in seven years!

And that's when the screaming begins.

CPSIA information can be obtained
at www.ICGtesting.com
Printed in the USA
LVOW11s0130010817
543282LV00004B/331/P

9 780988 003026